Dining Space
宴遇

黄滢 主编

江苏科学技术出版社

西方情调 / Western Flavor

福餐厅	006	Forge
Haleakala夏威夷餐厅	016	Haleakala Hawaii Restaurant
首席内阁	028	Chief Restaurant
名利场餐厅	042	Vanity
赛特夜总会	056	Set Nightclub
Kwint餐厅	064	Kwint Restaurant
摩尔花园	070	Moorish Garden
路易酒吧	084	Louis Bar-Lounge
奶奶家餐厅	090	La Nonna

现代创意 / Contemporary Creativity

Twister餐厅	100	Twister Restaurant
DN创意餐厅	106	DN Innovation Restaurant
Sun Alpina Kashimayari餐厅	112	Sun Alpina Kashimayari Restaurant
大妙火锅餐厅	118	Da Miao Hotpot Restaurant
柒公名豪餐厅	126	Qigong Minghao Restaurant
芸香小酒馆	138	Rue d'or
阿诺洛杉矶店	144	Nobu Los Angeles
阿诺达拉斯店	150	Nobu Dallas
唐宫海鲜舫	154	Tang Gong Seafood Restaurant
银河宾馆唐宫海鲜舫	162	Tang Gong Seafood Restaurant
酒醉俱乐部	168	Press Club
Jaga推广车	174	Jaga Experience Truck

Contents 目录

爵士酒吧	178 Brown Sugar	意粉屋	266 The Spaghetti House
Room 18+18 Love酒吧	186 Room 18+18 Love	怡亨酒店明园餐厅	272 Yi Heng Ming Garden Restaurant
拉维达餐厅	196 La Veduta	中意汇意式餐厅	278 River Club Restaurant
阿诺餐饮威基店	200 Nobu Waikiki	Meltino咖啡馆	284 Meltino Bar & Lounge
阿诺餐饮莫斯科店	206 Nobu Moscow	天筑餐馆	292 El Japonez
墨西哥酒吧	214 Nisha Acapulco Bar Lounge	Salon Des Saluts酒吧	298 Salon Des Saluts
桑普酒吧	228 Thumper	Shook!餐厅	302 Shook!
赤鬼炙烧牛排崇德店	230 Oni Boy Steak Chongde Shop	Affranchir餐厅	308 Affranchir
赤鬼炙烧牛排公益店	240 Oni Boy Steak Gongyi Shop	The Japonais餐厅	312 The Japonais
前门JM餐厅	246 Capital M	大卫迈尔斯咖啡厅	318 Davis Coffee
牛排城四望亭店	252 Steak City Siwangting Shop		
米斯特比萨店	258 Mr. Pizza		

Western Flavor
西方情调

福餐厅
Forge

▶ 设计公司：FFD设计
▶ 设计师：弗朗索瓦·弗罗沙德
▶ Design Company: FFD
▶ Designer: François Frossard

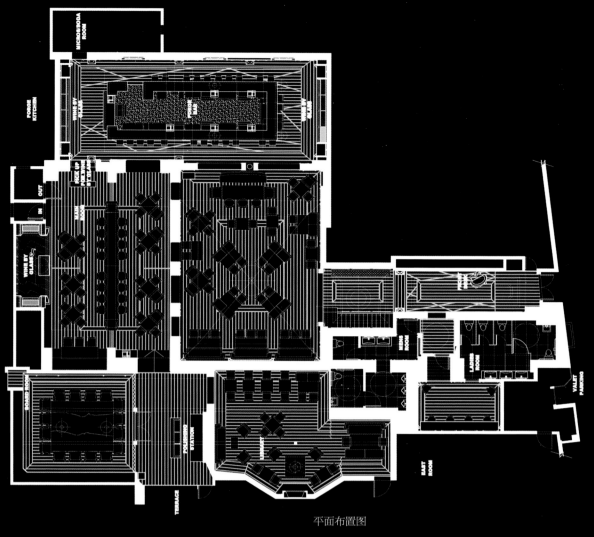

平面布置图
Layout Plan

尽管极富挑战，福餐厅最终还是以完美无瑕的品质赢得业主及其来往客户的厚爱，其中包括已经卸任或在任的各级政要及诸多社会人士。本案设计极具纪念意义。在新经济形势下的美国，设计师弗朗索瓦因锐意进取、突破陈规而声誉鹊起。他给本案营造出乐观向上的气氛，堪称典范。

本案在2010年3月31日揭开其神秘面纱。装修新颖、现代、讲究，兼收并蓄。传统的深色木材、主餐区宝塔型的大吊灯、粉红色的枝型大吊灯、彩色玻璃壁画、3.7m高的维多利亚式壁炉和令人惊艳的艺术品随处可见。

整个空间就像一个装饰艺术的大集合。建筑艺术、家具艺术、雕塑艺术、玻璃艺术、绘画艺术、波普艺术充满了每个空间，让人目不暇接。

入口处，一个如冰淇淋融化般的紫色接待台给人以无限惊喜。当你穿过黑色的走道，来到近两层楼高的白色大厅里时，眼前豁然开朗，白色的墙上各种建筑纹样的刻绘显示出空间的精致。最让人惊叹的还是那些在教堂里才能看到的华美雕花玻璃彩窗，绚丽的色彩、精美的图样让人眼花缭乱。

主餐区原计划要摆放100多个座位，后此计划遭到设计师的否定。设计师对空间重新进行规划，减少座位，采用更为自由的组合排序方式，因此主餐区显得开阔、畅通，让就餐者享受到更为轻松而优越的环境。独立包间内，实木大桌与白色高背椅的搭配新颖素雅，令人眼前一亮。移步入内，这里的空间被分为多个不同风格的进餐区，造型华美，组合方式多样，可以满足不同人群的进餐需求。长桌用餐厅当中的一条实木的长条桌板，可容纳近二十人同桌用餐，天花上的彩色蚀刻玻璃花吊顶让人倍感尊贵。

吧台区中的超大的吧台面可以容纳更多的客人在这里品酒聊天。两盏粉红色的枝型工艺大吊灯将奇异与华美融为一体，营造出独特的氛围感受。

本案连卫生间的设计也显得与众不同，超长幅面的人体艺术画作连成一体，坦露出摩登都市里的性感与自我格调。

本案的家具软装全部量身定做，而且大多数由空间设计师弗朗索瓦自行设计，风格多样，充满想象力，让整个空间都流露出高雅华贵和卓尔不群的气质。

Although Forge was a big challenge, it wins its reputation because of its perfect quality. Those who love this design include current and former heads of state and celebrities. In the new American economy climate, François gains high reputation with keen determination and without conversation, creating an optimistic atmosphere and a good example.

The restaurant's renovation featuring a new, more modern, eclectic design and décor was unveiled on March 31, 2010 by François Frossard. The restaurant's dark wood, pagoda-shaped chandeliers, big pink chandeliers, stained-glass murals, 3.7-meter-high Victorian fireplace and gilded-framed artworks can be seen everywhere.

The whole room is a big collection for various decorative arts like architecture, furniture, sculpture, glass, drawing and pop. They meet your eyes everywhere in the whole room.

A purple service desk like a melting ice cream is placed in the entrance. Going through the dark corridor, you come to the big white hall with delicate patterns carved on the walls. What surprise you most are the majestic and beautiful glass windows with brilliant colors and delicate patterns, which would win much admiration once one met it.

More than 100 seats in the main dining area were eliminated to create an open floor. A new plan for the space was thus made with the reduction of seats and new ways of combination. Therefore the space looks big and smooth, making an excellent environment for diners. There're big wooden desks and white chairs with high back. They look simple and elegant, giving you a big surprise. Inside further, the space has been designed into dining areas of different styles and combinations to meet customers' needs. The long solid wood table allows 20 people to dine at the same time. The colourful suspended glass ceiling gives an air of elegance.

In the bar area, the huge bar desk allows lots of people to enjoy drinking and talking together. Two pink chandeliers bursting with uniqueness and brilliance create a special atmosphere.

Even the design of washroom is unique with a huge painting of human body, showing the sexy character and personal style.

All of the stylish and imaginative furniture and instillations are tailor-made, most of which are designed by Frossard. They distinguish this project with uniqueness and elegance.

Haleakala夏威夷餐厅
Haleakala Hawaii Restaurant

- 设计公司：天坊室内计划
- 设计师：张清平
- 面积：307m²
- 主要材料：人工茅草、陶砖、奈米漆、不锈钢铁件、台湾杉木、玻璃纤维强化塑胶

- Design Company: Tianfang Interior Design
- Designer: Zhang Qingping
- Area: 307m²
- Material: artificial grass, ceramic tile, Nano coating, stainless steel iron, fir, glass fiber reinforced plastic

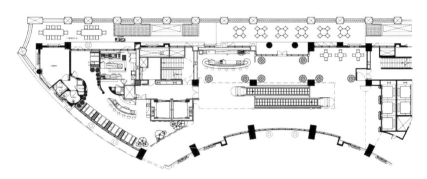

平面布置图
Layout Plan

想象着，草裙舞者在沙滩上轻摇裙摆，四弦琴"Ukulele"奏响的柔和曲调随着海风飘散在棕榈树间。在自然的见证下，人们许诺爱的誓言；夜晚时刻，在灯光和火把映照下，人们携手在闪闪发光的细致沙滩上漫步，享受唯美而又浑然天成的浪漫氛围，Haleakala夏威夷餐厅正是这样充满浪漫风情的天堂！

为求在流动的空间内为宾客们创造一种愉悦交谈的氛围，设计师将业主十分钟爱的夏威夷风格，原汁原味地搬进空间中，打造出海岛休闲气氛十分浓郁的餐饮天堂。

原始狭长的空间结构带来了设计规划上的局限与挑战，设计师以夏威夷特有的挑高斜屋顶独栋度假木屋作为空间主题，搭配最有效率的环绕罗列式动线，将原始质感的木材、砂岩表面、浮雕、手工编织等建材细腻地铺陈，将自然、热情海洋风刻画进空间的每一处。

夏威夷海滩亲切的小酒吧，搭配环绕格局，加上科技LED灯光重新诠释，成为布局的焦点。纯正夏威夷式的休闲加上科技浪漫，营造出整体餐厅星光海滩的绝美夜色情调。在这里用餐让人充满想象，你仿佛还可听到阵阵海潮欢乐的声音。自然热情的色调为人们带来快乐愉悦的用餐气氛。巨型扶桑花、茅草、贝壳、壁画等以热带元素为主题的装饰和随着空间错落有致地放置的水果塔、饰品相映成趣，传递着岛国风情，让客人有仿佛置身夏威夷热情异国的感受。的确，这样的气氛可以让人从都市的繁忙中抽离，在此多待上一整个下午，都不觉得腻。

设计是一个过程，而不是结果，因此设计师所融入的是主人本身的创意及构思，而不只是一味地表达设计师的观念。餐厅设计的原本构想是要留下更多的互动与想象空间，让体验的人亲自发掘其独特之处，我相信这里处处充满惊喜，一定能引起宾客们的热烈讨论，成为当天的话题，而餐厅本身也绝对不只是一个房子或是一个空间而已，它将成为一个开端，在餐厅主人的爱与关注之下，伴随着人们的互动发展，延展为一个具有延续性和生命感的故事空间。新奇有趣的故事情节和热情浪漫的主题，将在这里不断地上演延续。

Imagine that the hula skirts are dancing on the beach, while the Ukulele is playing soft music. At the witness of nature, people make the vows of love, walking on the beach in the light of firebrand hand in hand and creating the romantic atmosphere of beautiful nature. Haleakala Hawaiian-style restaurant is such a romantic paradise you will choose.

For creating more enjoyable time for consumers to chat, designers create an original taste of Hawaii style, making a casual dining space which the owners like very much.

The original narrow spatial structure has limited the planning of the space. But the designers make the Hawaii-specific single-family vacation cabin with a highly oblique roof as a space theme. Using moving lines, texture wood, sandstone, relief sculpture, hand knitting, and other building materials, the designer lays the delicate, natural and warm ocean into the space of every description.

Friendly little bar on the Hawaii Beach designed with circling layout, and coupled with LED lighting has become the focus of the space. Pure Hawaiian-style leisure plus technological romance can create a romantic atmosphere of night stars and stunning beach. Having dinner here as if you can hear the voice of tide; the natural warm tone creates a happy and pleasant dining atmosphere; giant hibiscus, grass, shells, paintings of tropical atmosphere create an island style restaurant; orderly scattering fruit towers, accessories make guests feel like being exposed to warm and exotic Hawaii. Indeed, such an atmosphere can make people abstracted from the busy city, and would like to stay longer.

Design is a process rather than a result. Therefore, integration of designers is the creativity and ideas of the owner rather than relying on the designers only. The original intention is to leave more room for interaction and imagination, and to make the guests discover the uniqueness of the space. Personally, I believe that the space is full of surprise, which can cause some heated discussion for the day. And the restaurant itself is definitely not just a house or a room, but a start, and with the restaurant owner's love and perfusion, it will be the interactive development and extension of the continuity of life story space room. Happy, warm, and romantic story will go on here.

首席内阁
Chief Restaurant

- 设计公司：成都市葵美树环境艺术设计有限公司
- 设计师：彭宇、张中源
- 面积：1 600m²
- Design Company: AOI Kankyo Design & Art Co., Ltd.
- Designer: Peng Yu, Zhang Zhongyuan
- Area: 1,600m²

首席内阁坐落于成都双楠片区内，由一售楼部改建而成，主营川粤系中餐。该店一改以往的传统中式风格，结合原建筑风格特点及对市场的分析，力求塑造出一个具有中东伊斯兰异域风情的用餐环境。

入口处，由原建筑门廊立面改造成的一个高大的弧形拱门给人以视觉上的冲击，远远地就引导着食客的美食神经。进入接待厅，高高的穹顶和弧形的窗洞，配以石材的柱子及经典的地面拼花，浓重的伊斯兰气息扑面而来。转入中庭，高达8m的浮雕墙上布满了图案，正前方放置的一盏流光溢彩的玻璃装饰灯映入眼帘，左右两边被拉长设计到直通屋顶的柱子，挺拔的柱身下面的基座被花瓣造型的浅水池围绕。整个中庭由奶油米黄石材、高密度雕花板加白色钢琴漆、艺术涂料及磨边茶镜等形成一个浅米色的高调色彩空间，把大气和精致体现得淋漓尽致。

中庭右边是明亮洁净的明档区；左边是六个带有户外花园的中型包间，户外花园可直通餐厅后花园，其中四个是可打开来举行小型宴会的四连包。

由左边的楼梯间上到二楼，蓝色和绿色两种基调的十个包间再次让食客仿佛来到中东伊斯兰的海边。其他五间夹层的处理，既增加了空间的造型种类，又带来了更多的休闲功能。

灯光系统和软装饰的搭配是本案设计中的点睛之笔。可按不同要求调成不同情景模式的照明，这样既节约了能源，又丰富了视觉。而餐具上的主题图案、精致的桌布、华美的灯具、别具一格的包间铭牌、木雕装饰品等，每一处都透露着设计师的精妙心思。

Located in Shuangnan Community, Chengdu, the restaurant is converted from a sales department, specializing in Sichuan and Guangdong food. The building volume is a place breaking away from buildings of the traditional Chinese style, and combines features of the original architectural style, shaping a Middle East Islamic exotic dining environment after the market analysis.

At the entrance stands a tall arched door transferred from the original façade, giving a visual impact, stimulating and drawing eaters. Into the reception hall comes a strong Islamic flavor out of the high dome and the curved window opening, accompanied by stone pillars and a classic parquet floor. Relief sculpture walls of up to eight meters in the atrium are covered with patterns, and a glass decorative light comes into view, while both sides are stretched to the roof pillars whose bases are walled with a shallow pool in a flower shape. The atrium brings out the exquisite atmosphere to the full, with cream beige stones, and highly carved plates coated in white piano paint and art, showing its delicacy and magnificence.

The right middle atrium is a bright and clean area while the left lined with six

medium-sized boxes in the arms of an outdoor garden; the outdoor garden leads to the back garden and four of the boxes can be used for small banquets.

The left staircase leads to the second floor, on which another ten boxes of blue and green allow an Islamic experience of the Middle East. Another five in the mezzanine adds more leisure function.

Lighting system and soft decoration make the best point adaptable to different requirements, saving energy and enriching vision. The theme patterns of tableware, the fine tablecloths, the beautiful lamps, the unique nameplates of boxes, the wooden decorations and so on, everywhere, reveal the subtleness of the designers' ideas and thoughts.

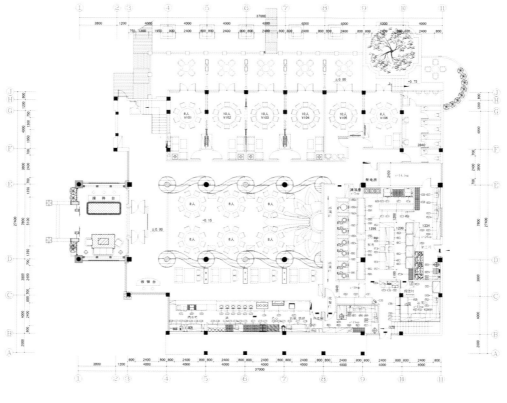

一层平面图

First Floor Plan

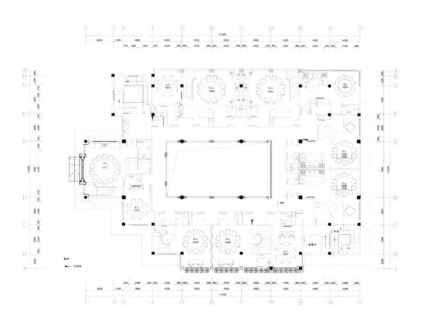

二层平面图

Second Floor Plan

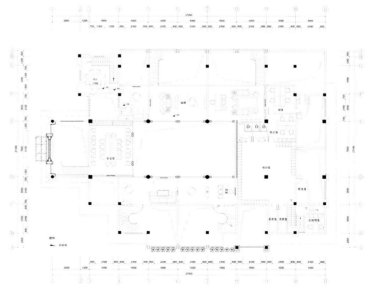

夹层平面图

Mezzanine Plan

名利场餐厅
Vanity

- 设计公司：重量设计
- 设计师：查尔斯、米里亚姆、桂贝斯
- 面积：1 260m²
- Design Company: Mr. Important Design
- Designer: Charles Doell, Miriam Marchevsky, Gui Bez
- Area: 1, 260m²

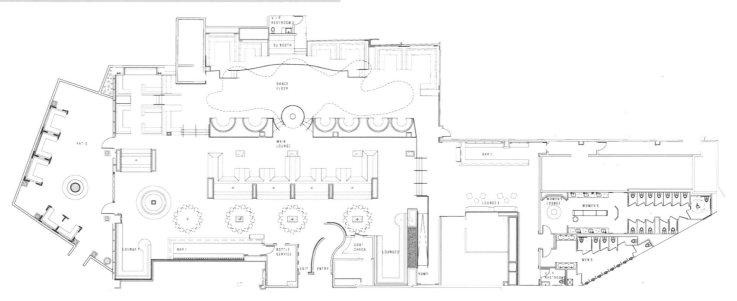

平面布置图
Layout Plan

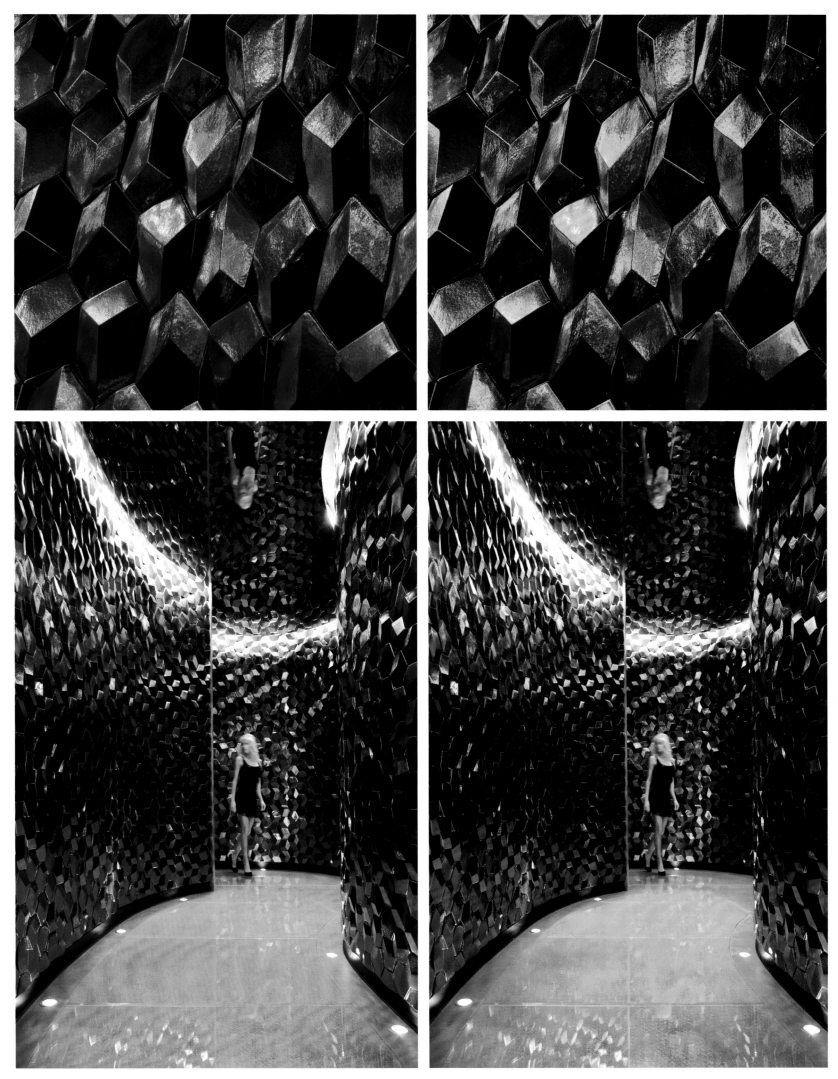

名利场如同一个折中处理的珠宝盒，具有极强的层次感。多面体的光结构，紊乱中依然有序，有序中却似无章。柔软的切面天鹅绒、绸缎、挂毯映衬着空间的奢华气质。珠宝、金饰、铜饰、古镜和擦铜、镀铬和黑等各色搭配相融。晶莹的珍珠、手工切割的水晶比比皆是。闪闪发光的表面纹理、蜂窝处理的表面漫游于空间，为内部空间塑造了一个有机的背景。

简单地说，名利场名副其实，正如珠宝盒给人的感觉一样，可以激起对名利的种种幻想。不同形式的切面、家具应用于空间内外，20世纪70年代、80年代的主观想象尽在其中。电脑控制的吊灯悬于中央，书写着曾经的传承、现代的时尚。

空间中火树银花，宛如街市。两万电脑控制LED微片、两万水晶共同铸就了灯火辉煌。庞大的灯组，气势恢宏，似玻璃地板中钻地而出，在高达6m的空中迸发，在舞池中蜿蜒起伏。倏尔视频传送，倏尔气氛活泼，倏尔色彩变幻，倏尔信息传递——动力因素，尽在程序。高科技因子，倍添空间内涵。叶连娜摄影作品也于空间大量使用。上世纪40年代风靡一时的《沉思女儿》摄影作品悬于其中，或于镜后，或放大6m置于舞池，或于浴室入口，或于次休息区。

女士专用休息室独具特色。全高的集层玻璃墙体，镶嵌时尚摄影师乌尔里希之大作，彰显女性高贵、典雅之特质。一端，金色的眼睛脉脉含情；另一端，秀美的嘴唇金光闪闪。两者之间嵌以白云，间以浅绿色、白色玻璃球，装点河山。独具特色的洗手盘乃亚米海因之杰作。特配的化妆镜，银光闪烁；柔软的树梅色发髻，高高耸立，彰显着洗手盆与众不同的身份。两个三面落地镜点燃空间激情，提升空间繁华、奢侈质感。休息室向里即是卫生间，呈簇状分布。其印刷图形，一横一竖，一笔一划，尽显超现实主义的效果。

Vanity is an eclectically layered club like a jewel. An organized chaos of multi-faceted forms is filled with soft cut velvets, silk fabric and rich tapestries. It is a mix of saturated jewel, gold bronze, antique mirror, rubbed brass and black chrome. Pearls and hand-cut crystals abound. Gleaming textures and honeycombed surfaces wander throughout the building, providing an organic backdrop for the club layers of reflective surfaces and parallax views.

Simply, Vanity is a sensitive and controlled playground, precisely like the feel from the jewel box causing fantasy to vanity and fame. To this jewelry based foundation we freely mixed forms of surface and furnishings associating with a variety of time periods but largely evoking from 70's and 80's. The space inherits the tradition and shows its current fashion with computer controlled chandeliers hanging in the air.

This one element is composed of 20,000 controlled LEDs and 20,000 crystals. It rises out of a glass floor and bursts up 20 feet high where it spreads and undulates over the entire expanse of the dance floor. It is capable of streaming video, animation, changing colors, quick message transmission. High technology and photographic works from Ye Lianna add much elegance to it. The photographic works of *Meditation Daughter* in 1940s were among it, some behind the mirror, some in the dance floor, some at the entrance of bathroom, other in the sub-resting area.

The other area of special character is the women only lounge which features full height walls of laminated glass with artwork produced by fashion photographer Miles Ulrich. One side has a beautiful golden eye and the opposite has lips dripping with gold. Between the two are white clouds as well as light green glass balls and white glass balls, rivers and mountains. The vanity area features sinks designed by Jaime Hayon, with individually vanity mirror and soft raspberry colored chenille poufs for each sink area. Two sets of 3-way floor to ceiling mirrors add the complexity and luxury of the space. The rear area of the lounge contains the bathroom stalls, which are "tufted" in a printed graphic for a surreal effect.

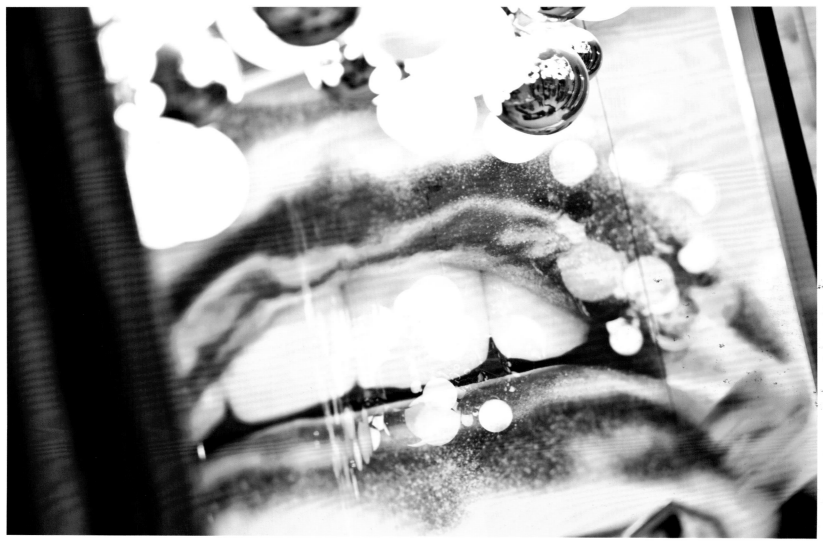

赛特夜总会
Set Nightclub

- 设计公司：FFD设计
- 设计师：弗朗索瓦·弗罗沙德
- Design Company: FFD
- Designer: François Frossard

赛特夜总会位于迈阿密海滩林肯路，原为剧院。整改后的酒吧引领"欧洲劲舞潮流"，空间气势雄伟，但毫不做作；设计精致，丝毫看不出前世的痕迹。壁炉点燃空间激情，电梯玻璃空间晶莹发亮，二者相互映照，沟通两个原本完全不同层次的空间。所有家具铺陈，全部用意大利著名时装品牌璞琪面料装饰。吊灯理念源于玻璃大师奇胡利。因为本案空间的设计，弗朗索瓦收获了"最佳设计奖"。

SET Nightclub, a transformed bar located on Lincoln Road, Miami Beach, originally was a theater. After its renovation, it leads the fashion of vigorous European dance. Although SET is majestic, it is free of pretension and a shadow of the building's former body. Frossard installed a fireplace and glass elevators that connect the two-leveled venue. Each piece of furniture is covered by luxurious Pucci fabrics. The concept of chandeliers was inspired by Chilhuly. Francois Frossard won a "Best Designer Award" for the club.

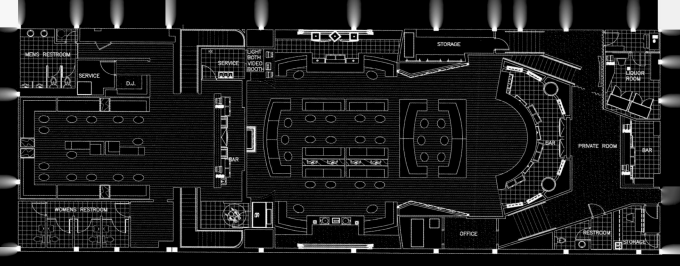

二层平面图
Second Floor Plan

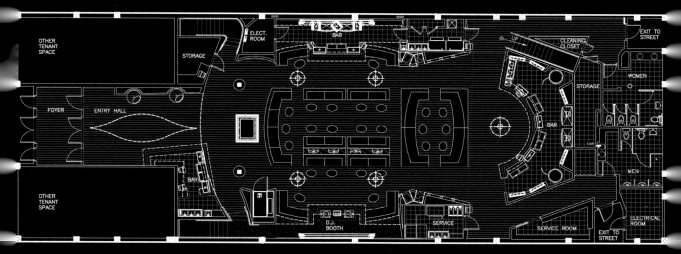

一层平面图
First Floor Plan

Kwint餐厅
Kwint Restaurant

▶ 设计公司：SAQ设计
▶ 面积：250m²
▶ Design Company: SAQ Design
▶ Area: 250m²

平面布置图
Layout Plan

Kwint餐厅坐落在布鲁塞尔市中心，朝向布鲁塞尔著名的Art-Mount。往来客人多出入于新开张的布鲁塞尔广场会议中心。本案前身为一废弃的顶级商场。此次翻修可谓对办公用房建筑遗产的尊重。空间设计风格极简，气氛独特，用材有限，犹如一缕春风，在业内令人耳目一新。

边墙以软体包装。Arne Quinze创作的30m长雕塑在餐桌上方徘徊，消失至空间的尽头，如同一个鲜活的有机体，护卫着就座其中的客人，强化着一种团聚的气氛。

软包墙其实是一个完美的声学系统，对周围环境的声音具有很好的吸收功能。厨房、采暖、通风、电力装置等基本技术要素、基础设施一应俱全；卫生间、休息室依次铺排。无规律的点装，如雕塑移动影化而成，强化墙体美观，尽显个性。

主色调统一和谐，雕塑表皮光洁、富有铜色质感，边墙开口裂纹处理，悄无声息地应对着透过大面开窗而入的自然光线。

夏日里，轻启折叠窗，布鲁塞尔古老的市中心一如溪水，轻快地流淌。

Located in the center of Brussels, facing the Art-Mount, the restaurant Kwint is designed as a central meeting spot for the newly opened Brussels Square Conference Center. The renovation of the formerly abandoned top-arcades in which the restaurant took its premises is the respect of the architectural heritage. It's simple with a unique atmosphere and a limited amount of materials, giving a new look for the insiders.

The length of the space is emphasized by both the padded side-wall and the sculpture hovering over the dining tables. This 30-meter-long sculpture is a creation of Arne Quinze and erupts from the bar at the end of the room almost like a living

and articulated organism. Floating above, the installation gives the seated customers a sense of protection and intensifies the notion of gathering.

The upholstered wall not only functions as a perfect acoustic absorbent for the ambient conversations, it houses all the essentially technical elements and infrastructure: the heating and ventilation system, the electrical installations, but also annexes rooms such as the kitchen, the toilets, and the lounge. Decorations of irregular arrangement highlight its beauty and character.

One color prevail, and the texture of glossy copper presents on the sculpture's skin as well as on the crackled surface of the side-wall openings, reacting quietly with the incoming daylight offered by the generous large-front windows.

In summer, the folding windows are opened up, with a view of ancient Brussels city center like a stream, lively.

摩尔花园
Moorish Garden

- 设计公司：成舍企业股份有限公司
- 设计师：周志荣
- 摄影：成舍企业股份有限公司
- 撰文：林雅玲
- Design Company: Fullhouse Interior Design
- Designer: Zhou Zhirong
- Photograph: Fullhouse Interior Design
- Text: Lin Yaling

小时候人人都喜欢天方夜谭的故事，明明那生成的背景和异国文明根本遥不可及，但那些出奇鲜明的影像——金色圆顶、神秘大沙漠、曼妙缤纷的丝缎、沙龙，总是在梦境里轻轻招手，让无数崇拜浪漫的灵魂随之轻盈飞舞，神游那些充满华丽冒险的精灵国度。

然而像这样只适合出现在梦境里的异国建筑，现在居然活生生地矗立在中国台湾亮丽的蓝天之下，这是由成舍公司的周志荣设计师负责指导的主题式餐厅：摩尔花园。建筑融合了温暖富足的北非色调与南欧西班牙等地极具辨识度的民族风格，精心打造出一处绝无仅有的手工梦幻城堡。

灵感：北非摩尔人的珍贵遗产

西班牙与非洲仅以15km宽的直布罗陀海峡相隔。中世纪以来，西班牙被北非摩尔民族统治长达700年之久，因此在建筑风情上洋溢出浓郁的伊斯兰色彩似乎是理所当然。尤其是在阳光闪耀的安达鲁西亚地区，多数象征性建

筑均混合了天然木材。闪耀的黄金色与黄铜材质，搭配赤陶与赤土瓷砖等特殊材料，将以土质特征为主的染料运用得十分彻底。换句话说，东西方文明相互激荡所产生的文明火花，不仅在当时改变了人们的生活形态，锻造出千变万化的地貌，更在数百年后给现代设计师提供取之不尽的制作灵感。

反观这些坐落于中国台湾的奇妙建筑，原汁原味复刻自遥远的真神属地，其间也穿插不少高迪的活泼创意。虽然与周边民情大相径庭，却忠实凝聚了业主与设计者双方共同的奇想与勇气，而且也为国人打开一扇观察世界的美丽之窗。总建筑面积达300m²，仅在四边局部塔楼有复杂楼层设计，大致以回形格局加上并列回廊环抱优美的喷泉花园为主体。在绝美的夜色、星光、灯光等的辉映之下，不知道引诱多少双手不自觉地猛按快门，这完全是被这遗世独立的超梦幻风情所深深吸引。

材质运用成就建筑特色

无论是砖砌庄园般的结构主体、拱顶尖塔的四边角楼，还是宛如一千零一夜场景中的拱柱回廊设计，由内到外所见都是鲜明的大地色彩，缤纷的马赛克拼贴，多彩多姿的铸铁玻璃灯饰。随处可见设计师者对于情景细节的精益求精。

以华丽见长的摩尔式建筑举世闻名，不同于罗马风格的沉重、浑圆。摩尔人擅长的数学计算，促使他们的建筑倍显轻巧而通风，这样的建筑特色在摩尔花园中同样清晰，尤其是充满伊斯兰色彩的庭院造景。园内有种植的中东海枣之类的特色植物，搭配比例精准的喷泉、水池。即使烈日当空，只要一进入这样的建筑体中，清凉的微风立即迎面而来。倒影在水面的美丽廊柱与细丝般序列圆柱上，承接着雕刻精美的装饰拱顶，这是伊斯兰建筑最鲜明的特色之一，而墙壁看似艳丽的花砖镶嵌，实际上是以水泥多次拓模、上

色、刷洗而创造出的高度工艺结晶。除此之外，在建筑周边饶富情趣的塔楼群也让人一见难忘。利用玻璃纤维雕塑出的或尖或圆的塔楼构件，在强化安全性之后，分别上色涂装，有些外观是淡淡的鲜艳色彩，有些则费心地以多彩马赛克拼贴。在部分塔身还能见到设计师挥洒的灵感，将破损瓷杯、马克杯等共同架构出令人悠然神往的异国情境。

在伊斯兰宫殿大啖西班牙美食

走近粗犷圆石堆叠，展现眼前的是厚实却充满诱惑力的神秘塔楼入口和无法一窥堂内的阴暗甬道。甬道尽头明亮的光，以及写满时光秘密的砌砖步道，沿途的风景线透过建筑明与暗，开阔与局促的交叉变化，让每一颗前来寻幽访胜的心灵，都能深刻体验到光影的变化与曲径通幽的那种感动。而塔内宝蓝的穹顶，在涂料施作时刻意让色彩随意流动，在斑驳灯光辉映下，仿佛打开更深邃的想象夜空。

刚好洒进一室午后灿烂阳光的餐厅，里面摆设褚红色的质朴桌椅，搭配红色耐火砖与精致绝伦的水泥花窗，兼顾透光、透风、防雨、防盗等特征。美丽之余还很能彰显热情的北非色彩，这是结合北非特有的沙漠、岩石、泥、沙等天然景观颜色而成就的建筑新体验。享受极富特色的石板烧烤料理之余，千万别让自己的视线错过任何一处细节。取自北非土生植物的深红、靛蓝，加上黄铜色泽，带来一种无垠大地般的浩瀚感。在室内与户外的过渡地带，圆形的拱门与回廊前后衔接，经典的对开木窗或手工拓模的各种古典花窗，带动不停流动的风雨香气。或以垂直或以水平延展，和谐地表现各种建筑材料最登峰造极的艺术风貌，以及淋漓尽致的人文厚度。

Everyone, in childhood, loves a fairy tale, in which the background and the exotic culture are obviously too far to reach, but images within are surprisingly and frequently clear in the dream: golden domes, the mysterious deserts, graceful and colorful silk, and salons. They seem irresistible and attract little minds and young souls to explore areas and places full of adventure.

Dreamy and exotic architecture, however, now stands in the blue sky in Taiwan. That is Moorish Garden, a theme restaurant designed by Zhou Zhirong, a building that combines warm and rich tones of North Africa and highly recognizable style of well-known families in southern Europe, and a unique handmade dream castle that's rare to come by in other places.

Inspiration: Precious Heritage of the Moors in North Africa

Spain was historically separated by the Strait of Gibraltar measuring only 15 km wide. Since the Middle Ages, North Spain was under the rule of the Moors for up to seven hundred years, so Islam style architecture was natural, particularly in the Andalusia region, where most symbolic buildings were a mixture of natural wood, gold color with brass color, terracotta with terracotta tiles, a thorough dye dominated by soil characteristics. In other words, the interaction between East and West civilization, not only changed the way people lived and landscape where they existed, but provided inexhaustible inspiration for modern designers.

In Taiwan, China, the wonderful and authentic architecture finds its way and is interspersed with lively ideas advocated by Gaudi. Different from the conditions of the surroundings, but faithful to embody fantasy and courage preferred by the owner and designer, it opens a door for people to see the beauty of the world. The total area of more than 300 sq m, with expectation for the complex floor design on partial tower, is treated simply. When in light of night, star and lamp, spring garden in a fret pattern and lined with corridors, is certainly charming and attractive, with its super-fantastic style that is remote and captivating.

Achievements out of Materials

Bricks build up a manor-like main structure; the four-side turret is added with a spire and a dome. The cloisters and the arches are like those popular in The Arabian Nights. From inside to outside, all is mainly in earth hues. Additionally, colorful mosaics and cast glass lighting, hang everywhere in the air, revealing the expectation of the designer of making perfection out of the perfection.

Moorish architecture is known for its world-famous magnificence; unlike the Roman style that attaches importance on the heaviness and the perfectly roundness, and because the Moorish is good at math, it thus makes their buildings light and airy. Such features are as clear as possible in Moorish Garden, which houses Islamic garden landscaping of plants like those in the Middle East and fountains as well as pools that are in ratio of precision. Even with hot sun, anyone entering it can immediately generate a cool breeze. One of the most distinctive features of Islamic architecture is available, that pillars are mirrored in the water and fine serial columns are good supporters of beautifully carved vault. The mosaic tiles on the wall seems showy, and actually, are of cement but have undertaken several times modules, and polish. In addition, towers around also make a memorable impression. All tower members, either sharp or round, are made of glass fiber and painted after their security is strengthened. Some are brightly colored; some are coated in mosaics, and even some are images produced by broken cups, mugs, etc. Innovations like this are natural to make exotic and fascinating leisure that is no doubt to be charmed.

Islamic Palace and Spanish Food

Tower entrance stacked with rough boulders, solid and thick, are full of mystery and allure. The hall corridor is so dark that is not easy to get a glimpse, through which is a winding brick path filled with secrets of time till the end that suddenly becomes bright. Landscapes outside occasionally pierce into the dark and light, so that every heart can acquire a profound experience of changing light and shadow and the sentiment out of the exploration for those secrets. The sapphire tower dome has been desperately and freely painted, which, once cast on with mottled lights, seems to draw the curtain off the night sky.

When entering the restaurant, sunlight in the afternoon comes across the dining space with a beautiful passion of the dark red of rustic furnishings, and the red bricks as well as the exquisiteness of cement rose windows. The rose windows are transparent, airy and rain-proof. Such an aesthetic and practical combination is actually a new application of natural landscapes in North Africa, landscapes like desert, rocks, mud, and sand and so on. While enjoying the very unique stone barbecue, you can satisfy your eyes by appreciating details: with brass color, hues like deep red and indigo originating from North African native plants, bringing in an image of a vast land; for the transiting area of the indoor and outdoor, round arches and corridors make a good bridge role; all classical rose windows, wooden or hand-cast, are in the flowing air. Extending both vertically and horizontally, all various building materials in the peak of art are reached in harmony, showing a thorough image of humanity.

路易酒吧
Louis Bar-Lounge

▸ 设计公司：FFD设计
▸ 设计师：弗朗索瓦·弗罗沙德
▸ Design Company: FFD
▸ Designer: François Frossard

迈阿密有海滩，海滩上有个甘西沃特酒店，酒店里则有个路易酒吧。该酒吧受巴黎皇家宫殿启发。法式古董装饰以其绚丽的色彩、现代化的气韵为空间增添风采。法国国王肖像照引领墙体图案装饰，鸢尾纹、御用纹、豹纹，书写皇家风范。古董水晶灯具搭配现代照明。路易十五的椅子、王朝时期的家具、镶嵌式衣橱、镀金的销控台皆由弗朗索瓦设计。

Louis Bar-Lounge at the Gansevoort Hotel, Miami Beach—inspired by a royal Parisian palace, features an antique French décor modernized with brilliant colors and updated accents. Portraits of the pictures of French kings adorn the different patterned walls of fleur de lis, regal stripes and leopard prints. Vintage and crystal lighting fixtures are paired with modern lighting fixtures and Louis XV chairs. French throne-style period furniture, inlaid armoires and gilded consoles were all designed by Francois Frossard.

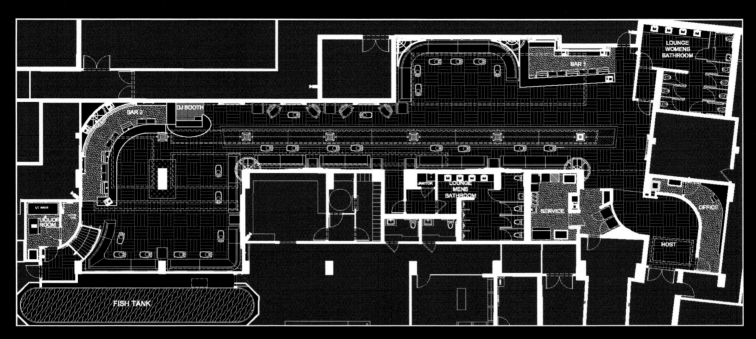

平面布置图
Layout Plan

奶奶家餐厅
La Nonna

▶▶ 设计公司：Cheremserrano设计
▶▶ 面积：200㎡
▶▶ Design Company: Cheremserrano Design
▶▶ Area: 200m²

奶奶家餐厅其实是意大利美食连锁。本案位于西班牙加纳利群岛拉康迪萨，是与DMG建筑师事务所合作的结果。餐厅以充分利用200m²的空间优势为布局起点，简约手笔融合餐厅环境，地产建材彰显地方特色，如黑石的地板、黑石的行进路线。墙壁、天花饰以特殊切割的镜面，扩大空间视觉感受。光影斑驳，洒下一地光华。家具风格典雅而简洁。中心放置了比萨饼炉，木吧台独立于中央。美酒享受，尽在其中。楼梯光洁明亮，服务等功能空间在光芒照耀中，自然攀援而上。夹层采光便利，摇椅、书籍和落地灯打造出温暖而生动活泼的家居气氛。楼梯处天花黑漆粉刷，彰显餐厅区的红砖设计。红砖中嵌入式照明，两相映照，设计美观，功能实用。

La Nonna is an Italian chain restaurant located in La Condesa. This project was made in collaboration with DMG architects. The design premise was to liberate the plan so the restaurant could take the most advantage of the 200m². The restaurant fuses its environment with simplicity. Incorporating the use of local materials, the floor is made of dark stones and is built onto the sidewalk, while the walls and ceiling are decorated with specially cutting mirrors. The mirror was placed in order to enlarge the space and create a fit of light and shadows. The furniture is elegant and simple. At the center is placed a bar with a pizza stove and a wood counter to sit and enjoy a nice wine. The services and the kitchen are placed on the second level connected with the ground floor by an elegantly illuminated staircase. When going down to the main level, the architects created a playful mezzanine that catches the eye of the visitor by letting light in and setting a swinging chair, some books and a floor lamp in the room. The ceiling in the stair and in the main circulations is painted black in order to emphasize the design of the red bricks placed in the restaurant area. The illumination is hidden within the bricks and creates a nice rhythm of light that accentuates that the architects are to integrate the architecture with the functional requirements and good appearance.

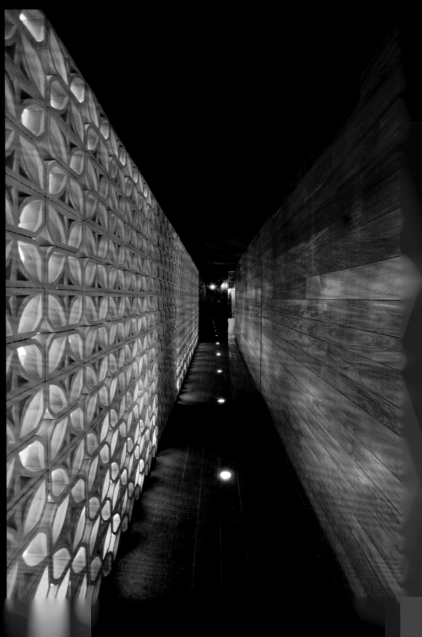

上层平面图
Upper Floor Plan

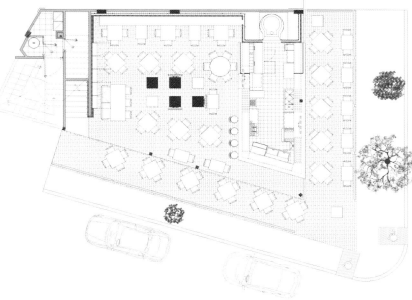

下层平面图
Lower Floor Plan

Contemporary Creativity
现代创意

Twister餐厅
Twister Restaurant

- 设计师：Sergey Makhno, Butenko Vasiliy
- 面积：421m²
- 主要材料：木材、混凝土、金属、大理石、塑料
- Designers: Sergey Makhno, Butenko Vasiliy
- Area: 421m²
- Material: wood, concrete, metal, marble, plastic

 Sergey Makhno

 Butenko Vasiliy

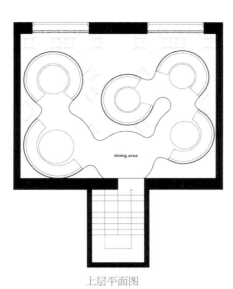
上层平面图
Upper Floor Plan

下层平面图
Lower Floor Plan

由Sergey Makhno 和 Butenko Vasiliy组成的设计团队完成了一家位于乌克兰首都基辅的餐厅室内设计。在这家餐厅享受鸡尾酒时你会觉得自己是一只雏鸟,或者是端坐在龙卷风顶上用餐呢!餐厅属于现代欧洲风格,设计的主要目的在于创造一个自然、现代、舒适的用餐空间。餐厅分为休息娱乐区和两层用餐区。

用餐区的设计灵感来自两个大自然现象:龙卷风和雨。位于上层的用餐空间以6个龙卷风形状的楼厅为特色,创造出一个由5个用餐小隔间组成的动态空间。餐厅墙壁是光滑的木条,悬挂在天花板上的灯模仿坠落的雨滴,营造出一个非常自然舒服的氛围。整个餐厅运用自然的米色、褐色、石榴红和棕色为主色调,使得整个空间显得自然平静。

休息区的墙面是用小树枝粘在一起做装饰,以营造鸟巢般温暖舒适的感觉。椅子的设计让人联想到松类的圆锥体和森林。整体设计给人们提供了一个放松及适合沉思的空间。

Design team of Sergey Makhno and Butenko Vasiliy have completed the interior for a restaurant in Kiev, where you will feel like a baby bird while drinking cocktail or having a dinner at the top of a tornado. This restaurant can be classed as modern European style and offers a molecular kitchen style dishes. The main aim while designing this restaurant space was to create an environment that is natural, modern and comfortable. This restaurant features two areas: a two-storied dining section and a relaxing bar area.

The two-storied dining section was inspired by two natural phenomena: tornado and rain. The space features six tornado shape balconies which create one dynamical upper zone with five dining cells. Restaurant walls were lined with smooth wooden boards. Ceiling lamps imitate rain drops falling from the sky, creating an atmosphere that is very natural and comfortable. Spaces of the restaurant are calm due to the natural tones which extend from the whole restaurant, including beige, ochre, and garnet, brown.

The wall in the relaxing area is decorated by wooden sticks stuck together among themselves. This wall decoration creates a feeling in a bird nest where you can feel warm and cozy. The design of armchair reminds us of coniferous cones and forests. The whole place is made for relaxation and contemplation.

DN创意餐厅
DN Innovation Restaurant

- 设计公司：咏翊设计股份有限公司
- 设计师：刘荣禄、刘沂娟
- 面积：379m²
- 主要材料：抛光石英砖、页岩、橡木染白、墨镜、镜面板、绷布、进口壁纸
- Design Company: Very Space International
- Designer: Louis Liou, Liu Yi-Juan
- Area: 379m²
- Material: polished porcelain tile, shale, oak dyed white, sunglass, mirror panel, stretch fabric, imported wallpaper

平面布置图
Layout Plan

DN 创意餐厅是一家西班牙美食料理餐厅，位于台北信义区。走入这个空间，行云流水般的线条，强烈的解构主义美学，让人几乎忘却了这原本是个近似方形的空间。空间面积并不大，379m²，却以流线型的隔断环围出一个洞穴般的空间。这个空间外观封闭，而内里敞开，空间的特色造型极具视觉冲击力，让人几乎忽略了实际空间面积的大小。

整个空间从一个巨大的平面柜台开始，逐渐转化成一个独特的现代主义雕塑空间——华丽流畅的线条、强烈的动感、跃动的张力，在空间中飞舞，展现有如来自外太空超现实的意境，让不规则的秩序与动态的平衡相互交融并存。经由这些不断变换的场景，我们得以自在地穿越、旅行——有如一道道令人惊艳的创意料理，循着感官的轨迹不断地感受着新的味蕾之旅。

整个空间采用了流线型分区，将餐厅内的流动空间连在一起，进入空间后每走一步都会带来不同的空间体验。全新的感官体验，揭示全新的征程，就好像主厨Daniel Negreira的美食所带来的那意想不到而又令人感动的体验。透明与不透明相结合，相互冲突却又奇迹般的和谐。

极具现代感，但风格不同的三个独立VIP包厢，可以满足不同顾客的隐私需要。其中一间可以观赏到整个餐厅的布局；另外一间，由帘子遮掩住；最后一个包厢则被完全隔离，提供一个完全隐秘的空间。一个设计别致的主餐桌摆放在整个餐厅的中心，为客人提供法国的克鲁格香槟。

天花板的部分设计使用了金属镜面，宛如一条静静流淌的天上河流，创造出一个极具梦幻的场景。黄色的灯光与西班牙朱红相映衬，创造了一个安静的氛围，突显了厨师的母国特征。具有特别纹理和色彩的家具都是定制的，使空间的所有元素融为一体，并使其成为空间延伸的一部分。

DN Innovation is designed for a fine Spanish restaurant located at the most sophisticated area of the Xinyi District in Taipei City, Taiwan province. Entering the space, the streamline and aesthetics will be well felt. Under this circumstance, one almost forgets that it's an approximately rectangular space. The size, 379m², is not very large. With the inswept walls, there is a cave-shaped space with an enclosed appearance and an open inside. Its type has a dramatic impact on people to make them ignore the real size of the space.

The whole space is gradually transformed from a big platform to a unique and modern sculptural space: spectacular and smooth lines, dynamic sense and flexible tension, which dance in the air, showing its super realistic state as if it's from the outer space. Irregular order and dynamic balance exist together. Through these changing views, we can enjoy a free trip.

Adopting streamlined partitions, the spaces within the restaurant are connected. After entering the space, you will get various experiences through different steps. With new feel, a new journey is revealed, which is the same to the taste of the food cooked by the chef Daniel Negreira. Transparent and opaque colors seem to conflict with each other, but actually, it's in harmony.

The three independent contemporary VIP boxes are designed to meet different privacy needs of the guests: one offering a complete view of the restaurant, one veiled by curtains and the last closed to offer total isolation. A special main dining table stands at the core of the restaurant, offering the excellent French Kruger's champagne.

Metallic mirror-like materials are partially used on the ceiling, creating illusions of a vivid river flowing in the heaven and presenting dreamlike images. The yellowish shade of lights juxtaposes with Spanish red tones, generating a tranquil ambience, exposing traits of the Chef's mother country. Furniture pieces with particular textures and colors are customized to integrate all elements into one, making them a part of the extension of the whole space.

Sun Alpina Kashimayari餐厅
Sun Alpina Kashimayari Restaurant

▶ 设计公司：The Wholedesign Icn
▶ 设计师：衫山敦彦
▶ Design Company: The Wholedesign Icn
▶ Designer: Atsuhiko Sugiyama

Sun Alpina Kashimayari 坐落在一个有丰厚积雪的滑雪度假胜地。其西部是日本北部阿尔卑斯著名的Kashimayari山峰，其东部是青木湖。

Sun Alpina Kashimayari餐厅外观是非常简约的现代风格，弧形的一面向着山景，可以最开阔的尺度观赏壮阔的雪域风光。令人印象深刻的是在一层走廊上那粗大的圆柱子，让人想起古时宫殿建筑采用的大圆柱，给人庄重稳固的感觉。

室内布置非常质朴舒适。大地色系的色彩在室内肆意铺陈，从地面、墙身到天顶，一气呵成，给人温暖的感觉。大量木材在墙身、地面、楼梯使用，让人想起童话中森林里的小木屋，随时等待疲惫的旅人在这里休憩。石块贴面的壁炉里柴火烧得旺旺的，暖洋洋的气氛在室内弥漫开来。

粗大圆柱设计也延伸到了室内，不过室内成排的柱子，每一根都是一个人也围抱不过来的粗大树桩子。不知道这些树桩子要经过多少年的风霜雪雨才能长成这样粗大，顶天立地支撑在室内，就像森林守护之神，让餐厅里的旅人可以安心休息与用餐。

餐厅每一层空间都是通透的大开间布局，长形小餐桌可以自由组合，无论是团队聚会，还是两个人私语，都可以各得其所，各得其乐。

在接待区，两根粗大木桩的顶上，一幅大大的雪山壁画，让人不仅可以仰望大自然的宏阔手笔，更可以感受到自然的神奇与无穷魅力。

Located in a place with a spectacular mountain view, Sun Alpina Kashimayari is a snow rich ski resort. In the west, you can see Mt. Kashimayari, a famous peak in the Northern Japanese Alps, and in the east, you can see Aoki Lake.

The appearance is of simple modern style with one arc side, facing the mountain, where you can enjoy the beautiful snow. What's more impressive are the huge columns on the first floor corridor, which seem like the ones used in ancient palaces, attaching an air of solemn to the space.

The interior is very simple and comfortable with the earth hues colored from the floor to the wall and the ceiling, creating a warm feeling. The massive use of wood makes it like a little wood house in the fairy tale in the forest awaiting weary passer-bys to take a good rest in it. The fire in the stone fireplace creates a warm atmosphere in the whole space.

The use of huge wood columns which must have taken many years to grow so big that a person can't hug it extends to the interior. Like the guardians of the forest, the huge columns make people feel at ease.

The whole space on each floor is very open and large. Rectangle small tables can be combined freely to meet the demands of people coming in group or in a private conversation.

Above the huge wood columns in the reception area is a large fresco of a snowy mountain, which shows the myth and charm of nature utterly and will win your admiration for its spectacular view when you are looking up at it.

大妙火锅餐厅
Da Miao Hotpot Restaurant

- 设计公司：成都市葵美树环境艺术设计有限公司
- 设计师：彭宇
- 面积：900m²
- 主要材料：花梨木、红色、米色、蓝色人造石
- Design Company: AOI Kankyo Design & Art Co., Ltd.
- Designer: Peng Yu
- Area: 900m²
- Material: rosewood, artificial stone

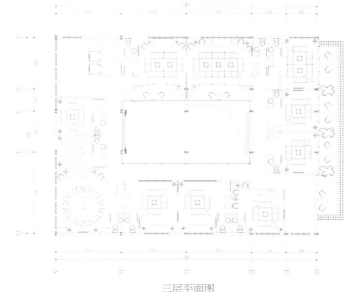

三层平面图
Third Floor Plan

本案位于2008年成都市重点恢复整理出来的原老少城中极具代表性的宽窄巷子景区，目前人气极旺，已成为在成都继锦里之后又一处旅游必到的人文景点。

大妙火锅餐厅在此将本土文化的特征继续发扬，空间构成中通过对原建筑的整理，形成完整的中庭空间，以利于舞台的表演。在将川剧脸谱中的代表图案做深化变形处理后，结合珐琅彩瓷器中的嵌丝技巧及色彩关系，重新进行设计并运用到空间造型中，形成既保留传统又不失当代审美的独特设计风格。

上层主要聚集了供接待、商务聚会的包房，这里准确地找出了可以借景的空间，将人文建筑特征极为浓厚的宽窄巷子景观引入室内，并在包房内设计中加入部分老少城的文化艺术片段，让人在一种淡雅的怀旧情结中体验大妙餐厅的个性文化。

二层平面图
Second Floor Plan

The restaurant located in the representative scenic zone wide-narrow lane, which was the key restoration of Chengdu city in 2008 and has now become one of the cultural scenic spots after Jin Li.

This project continues to develop their local culture by remaining the architectural structure of original architecture to form a complete and open lobby, providing the advantage for stage shows. The Sichuan opera face patterns have been redesigned with the skills of making decorative porcelain and its colors to be used in the design of the space and thus keep the unique modern aesthetic design styles within traditions.

On the upper floor are rooms for reception and commercial meetings. Here the traditional view of cultural architecture features of wide-narrow lane and some of the cultural art elements have been combined into the interior design, so that people can enjoy the restaurant's special culture with a nostalgic feeling.

柒公名豪餐厅
Qigong Minghao Restaurant

- 设计公司：浙江亚厦设计研究院
- 设计师：孙洪涛
- 参与设计师：蒋良军、胡杰涛、王蕊
- 主要材料：柚木、古堡灰大理石、爵士白大理石、黑金花大理石、古铜、景泰蓝、皮革
- Design Company: YASHA
- Designer: Sun Hongtao
- Assistant Designer: Jiang Liangjun, Hu Jietao, Wang Rui
- Material: teakwood, grey marble, jazz white marble, black gold flower marble, bronze, cloisonne, leather

一层平面图
First Floor Plan

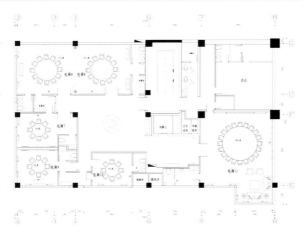

二层平面图
Second Floor Plan

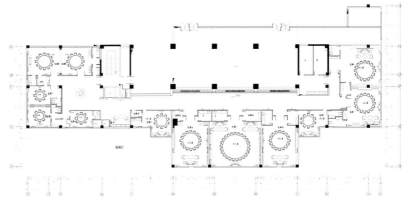

三层平面图
Third Floor Plan

梅，剪雪裁冰，一身傲骨；兰，空谷幽香，孤芳自赏；竹，筛风弄月，潇洒一生；菊，凌霜自行，不趋炎势。进入大厅，首先映入食客眼帘的正是这铜质雕刻的岁寒四友。梅兰竹菊，是君子的象征，蕴含中国人对最崇高的人格品性的赞美与向往。大凡生命和艺术上升到"境界"的层面，都致力于将有限的性格特质升华为永恒无限之美。

"祥云应早岁，瑞雪候初旬。"过厅的地面、墙面、顶面，都融入了祥云的元素。虽然表现方法有所不同，但其主要目的都是为了体现"渊源共生，和谐共融"的文化内涵。还有一些栅格、隔断，以及顶面的花格，都巧妙地点缀了空间。

中式的布局、器物，搭配欧式的家具、水晶灯，亦中亦西，抽象的水墨画，揉合了两者之间的矛盾与对比，使之衔接更为自然。

幽静儒雅的空间，极具翩翩君子之风，大气开阔的布局，更显雍荣气度。柒公名豪餐厅，尊在气度，豪在品味。

Plum blossom, with its essential character of pride, blossoms in the snowy season; orchid grows alone in the valley with a delicate aroma, reflecting the soul of independence and purity; bamboo thrives with its shadow swaying in the moonlight, representing the spirit of elegance without constrain; Chrysanthemum blooms in the extremely cold days, presenting dignity and integrity. Customers will first catch the sight of the above four durable plants of winter sculped by copper when entering the hall. Plum blossom, orchid, bamboo and chrysanthemum, as a symbol of the man of noble character, can convey the quality of loftness and nobility, which earns the admiration of Chinese. Generally, after rising to the philosophical level, the limited character of life and art will turn to eternal infinite beauty, which leads to sublimation.

"Cloud" is an important element which indicated a rich year in ancient China, therefore, Clouds elements are integrated into the design of floors, walls and ceilings through the hall, presenting the cultural concept of "origin and harmony" in various ways. Some grids, partition walls and the lattice on the top surface are applied into space.

The layout and wares of Chinese style, accompanied by furniture and crystal lamps of European style, represent the effect of Chinese and Western cultral interaction. Abstract ink painting helps to find a way out of conflict and contrast between Chinese and Western elements, of which the transition is made more natural and comfortable.

With space of tranquility and elegance, as well as the gorgeous opening layout of magnificence, the restaurant is renowned for dignity and taste.

芸香小酒馆
Rue d'or

- 设计公司：富魅公司
- 设计师：Yasumichi Morita
- 图片提供：Nacasa & Partners Inc.
- 面积：392m²
- Design Company: Glamorous Co., Ltd.
- Designer: Yasumichi Morita
- Photo Credit: Nacasa & Partners Inc.
- Area: 392m²

芸香小酒馆位于一楼，依走廊、人行道而立。如此的地理环境，要创造繁华、开放但又独具特色，融当地丰富的文化底蕴为一体的小酒馆，于本案是个不小的挑战。我们的设计理念是沿着双高窗户建造一堵图形陶瓷墙，并以开口处理。普通的法国家庭中和小酒馆里，那些摆满了厨房用品、书籍、灯罩、酒瓶、玻璃制品等的架子是其设计灵感的来源。一楼空间作休闲之用。曲面吧台，装饰着具有象征意义的巨型灯罩。后面是开放式厨房。厨房与它背后象征意义巨大的曲杆与灯罩柜台共同构成开放式厨房用餐空间。厨房如空间心脏。一楼空间上层舒适，气氛亲密，拥有极佳的用餐空间。闪闪发光的金属链饰、流苏和蚀刻铜镜提升了空间的奢华感与排场。巨大的设计墙镌刻着法国烹饪术语，其后隐藏的楼梯连接上下空间。青铜镜的装饰框景自然，来往行人如在画中。扶梯下望，蓦然发现，闪闪发光的窗帘，却是金属链条铸就的美丽。丰富的质感和色彩色调代表着精致的法兰西文化，为尊敬的来宾带来欢快的体验。

Located on the first floor along the public corridor and sidewalk, Rue d'or faces the challenge to create ambience of bustling French bistro that is open and unique while maintaining sophisticated elegance representing the rich heritage of St. Regis. A huge void along the double-height window is celebrated by iconic ceramic design wall. It is inspired by the shelf that can be found at a French household or the bistro both filled with miscellaneous kitchenware items, books, lamp shade, wine bottle or glass, etc. Cheerful experiences to look into this feature wall represent our design concept for Rue d'or. The first floor is a more casual dining space with open kitchen behind the curved bar counter with symbolic gigantic lamp shades. The bustling kitchen, which is the heart of the restaurant, will flow out throughout the floor. On the contrary, the upper floor is designed for cozier and more intimate dining and drinking spaces. Glittering metal chains, tassels, and design-etched bronze mirror adds elegant pomp to the space. Floors are connected by a staircase hidden behind the huge design wall that is engraved by French culinary terms. It has windows decorated by original bronze mirror frame that will make a scene of the guests going up and down as if they are in a picture frame. Also guests travelling by the staircase will find an interesting view of the restaurant below by seeing it through the frame before they find a glittering curtain made of beautiful metal chains. Rich texture or colors represent the exquisite French culture, while a series of cheerful design experiences are scattered throughout the space to entertain the guests.

阿诺洛杉矶店
Nobu Los Angeles

- 设计公司：罗克韦尔集团
- 摄影师：蒂姆·斯特里特·波特
- 面积：室内 471m²；庭院 146m²
- Design Company: Rockwell Group
- Photographer: Tim Street Porter
- Interior Area: 471m²; Outdoor Patio Area: 146m²

阿诺洛杉矶店位于一独立建筑，是洛杉矶西好莱坞橘园的前身。开张后的阿诺拥有全新的外观，独特的视觉元素，处处诠释着阿诺风情。

建筑立面生机无限。槽形墙、花岗岩质地和高贵的紫色调，都充满了生机。长短不一的蚁木梁柱、火树银花的秀木，二者依次交替而成。

手工制作的铁艺大门颇显欧陆风情。门楣上的一顶木质华盖和庭院里成排的木质板条，意蕴悠长。定制的门楣遮盖上布满了金属板条，凸显出现代空间的浓艳色彩。标志性的核桃木门自然而立。门内，当然是一片宁静。

酒吧休息室墙面粉饰，加上手工染制的茄紫色，延续着立面主题色调。其中部分还是罗克韦尔专为Maya Romanoff制作。乌木灰色的顶和青铜模制的条纹墙，浑然幻化成了树皮纹理的样子。后部的酒龛饰以咖啡色。定制的白茧吊灯悬在空中，照亮了主餐厅。在这里，人逢知己，尽可把酒言欢，加之高贵、清雅的环境，即使一个人，金樽也不至于空对。

休息室、主餐厅的木门如同景框，临近空间的景观一一映入眼帘，自然而不做作。门上悬挂的焦天鹅绒门帘全部是定制的，或桦色，或棕色，或乳白色，绘画着大大小小的桉树图纹。

阳光灿灿，白云飘飘，举目张望，恍然间，天窗下的餐厅中悠然立着一个庭院。界定墙简约、高雅，两边覆盖着水生风信子编织的复合板，自然的意味油然而生。沿着墙，桔黄、梅红掩映枝头，桉树的枝叶与花朵，娇艳欲滴。定睛一看，不禁惊叹于巧夺天工的手工艺术。三维的壁纸竟然是手绘、刺绣而成。数盏小灯簇拥着定制的吊灯。金属质感的编织，恰如灯罩，散发着富丽堂皇的感觉。闪闪银光，一张一合，如同天窗外点点繁星。

穿过庭院就是主餐厅，空间采用茄紫色系的色调。迈克尔·帕拉迪诺的兰花图跃然于锦旗之上，恰到好处地装点着背墙。背墙之下，悄然立着背光烧焦灰色的寿司吧。

主餐厅单独设有就餐露台，把单体建筑的优势发挥得淋漓尽致。露台加顶，尽显空间设计的人性化理念。四座席，采用耐用的户外材料制作。棕色皮浅色处理，深橙色的面料织以花卉图案，更有棕色的酒椰布应用其中。诸种元素悄悄地把空间融合在自然中。定制的蕉麻编织屏风如茧封一般，围合着座席区域。嵌地的照明，向空中挥洒着风情。定制的塞多纳红色成品道格拉斯菲尔烛台依墙而立，摇曳的红色，平空添了一丝浪漫，一抹红韵。

Nobu Los Angeles is located in a freestanding building that formerly housed L'Orangerie in the West Hollywood area of Los Angeles. The restaurant represents a fresh new look and unique visual elements, showing its local culture.

The facade of the building has been reinvigorated by painting the channeled granite wall in a bold purple, and building with a series of vertical Yimu beams of various

lengths and depths, along with trunk of big trees.

Guests proceed through a hand crafted iron gate of European style under a wood canopy, which together with rows of wooden boards show much excellence. The customized covers on the door, show the modern deep colors. The iconic Nobu wood doors of stacked walnut timbers naturally stand there, making a serene space in the interior.

The walls of the bar lounge are decorated in eggplant purple color by hand, continuing the main color of the façade, part of which are designed by Maya Romanoff. The bar itself has an ebonized ash top, and a bonded bronze bar wall with a striated pattern, giving rise to a shape of the tree skins. And the back of the bar is of coffee color. Custom-made white cocoon chandeliers are suspended in the air and illuminate the main dining room where you can enjoy your life with your best friends or you ownself.

The doors of the bar lounge and the dining room are like a view frame, from which the nearby views will all meet your eyes. The custom portieres of velvets are patterned with various eucalyptus patterns in the colors of white, brown or milky white.

With the brilliant sunshine and flying clouds, you suddenly see a courtyard in the center of the restaurant under the skylight. The wall that divides this room from the bar lounge is simple and elegant with eco-friendly water hyacinth weaved panels on both sides. The interior wall of the internal courtyard is covered with three-dimensional wallpaper which are painted by hands and embroidered with eucalyptus leaves and flowers, creating a sense of delicacy. Custom chandeliers crowded by many tiny lights are giving out a metal sense through the knits, which are just like lampshades. The glistening lamps are just like the stars in the sky.

Adjacent to the internal courtyard is the main dining room, which is entirely clad in eggplant colors. Two silk banners of Michael Palladino photographs of orchids adorn the back wall. The backlit scorched ash sushi bar is also located in this area.

Outside of the main dining room is a covered outdoor dining patio with four legs made of durable outdoor materials such as amber brown leather, a deep orange fabric with large brown woven flowers on it, and brown raffia fabric. Custom woven abaca screens envelop the sitting area like a cocoon, illuminated by the glowing lights built into the floor around them. Custom Sedona red finished Douglas Fir candlesticks stand against the walls. The flickering red gives rise to a little romance.

阿诺达拉斯店
Nobu Dallas

> 设计公司：罗克韦尔集团
> 摄影师：大卫·约瑟夫
> Design Company: Rockwell Group
> Photographer: David Joseph

阿诺达拉斯店开业于2005年。可谓是阿诺餐饮开拓德克萨斯市场的前站。用餐空间由罗克韦尔集团倾情设计。河石铸就的石墙，沧海横流。玛瑙质感的寿司酒吧，光彩照人。原始的质感让人浮想联翩，阿诺纽约总店的白桦风情如在眼前。但光芒之下，却无法掩盖德克萨斯的浓郁风情。或餐饮，或点心，或小啜，或休憩，透过天窗，阳光普照的空间中总有适合你的一处。

Nobu Dallas opened in 2005 is Nobu Restaurant outpost for the market of Texas. The Rockwell Group-designed dining room is of river stone wall while the onyx sushi bar is very reminiscent. The original texture Nobu reminds you of many things, sometimes as if the birch trees in its parent hotel Nobu Hotel are just before your eyes, showing its brilliant charisma. Nobu Dallas can be the destination for dinner, tea, drink and a nap. No matter who you are, there must be a place with the sunshine coming through the skylight you will like.

唐宫海鲜舫
Tang Gong Seafood Restaurant

- 设计公司：非常建筑
- 设计师：张永和
- 项目负责人：林宜萱
- 设计团队：于跃、吴瑕、王兆铭
- 面积：2 460m²
- 主要材料：竹、复合木板、水磨石
- Design Company: Atelier Feichang Jianzhu
- Designer: Zhang Yonghe
- Project Manager: Lin Yixuan
- Design Team: Yu Yue, Wu Xia, Wang Zhaoming
- Area: 2,460m²
- Material: bamboo, composite wood, terrazzo

餐厅位于杭州新城区大型商场的顶层，拥有将近9m的层高以及南侧开阔的视野。我们选用了复合的竹板作为主材料，成为强调传统与现代相结合的设计主题。

在大厅中，利用原有的层高优势，我们将部分包间悬吊于顶上，创造出高低层次的趣味性并丰富了空间的视觉感受，因为在原有的建筑条件下，大厅中心巨大的核心筒和侧边悬挑的半椭圆形体块使空间显得零碎杂乱。我们以一片用薄竹板编织、从墙面延伸至天花的巨大透空顶棚，将空间重新塑造。波浪起伏的竹顶棚，构筑了大厅里戏剧般的场景。而视线穿过透空的竹网，不仅保持了原有的层高优势，亦使得上下层有了微妙的互动关系。在原来的核心筒外，我们以透光竹板包覆四壁形成灯箱，则使得原本沉重的混凝土量体变为空间中轻盈的焦点。

入口门厅亦延续竹的主题。墙面覆以竹材，并顺应原有的墙体处理成如波浪状流动的弧面，除了与大厅的顶棚相呼应外，也具有空间导引的功能，并且令顾客一进入餐厅就有耳目一新的感受。

包间的设计则强调同中有异。一层的包间较大，从天花到墙面的折板和两侧镂花透光墙面是共同的基础语汇，但每间各自不同的折板角度和镂花图案，则使得各个房间有了彼此相异的面貌。南侧夹层上方的包间略小，借由特殊的曲面顶棚造型和简洁单纯的竹材墙面，营造出空间趣味并显其大方。至于作为空间重点的悬挂包间，以空桥和侧边走道连接，半透明的墙面形成隐约的内外关系，使人不论在其内或外都能产生特殊的空间体验。

在这次的设计中，我们希望借由对新型竹材的不同应用方式，能塑造出彼此相异却连贯、一体的空间感受；并在追求空间创意的同时，也保持对当地文化的尊重。

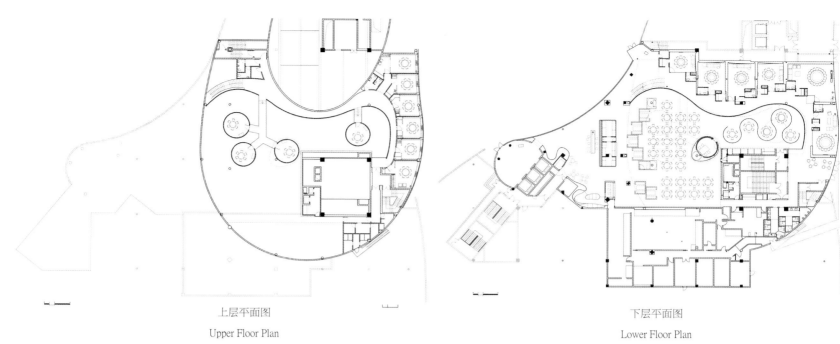

Upper Floor Plan

Lower Floor Plan

This restaurant is located on the top floor of a big business plaza in new town district of Hangzhou, 9 meters in height and facing the south open scenery. We use bamboo board as its main material to stress the design theme of combination of tradition and modernism.

In the hall, with the advantage of the height, we suspend some of the compartments on the top to create an interesting vision. We use thin bamboo boards to weave a huge plafond which extends from the wall surface to the ceiling to mould a new space, because in the original structure, there is a huge pillar standing in the middle of the hall with some semielliptic constructions on the side wall, which make the whole space look in disorder. The wave-like bamboo plafond makes a melodramatic scene for the space and keeps the advantage of the height while giving the up and down space a delicately interactive relation. As for the pillar, we cover it with bamboo board to shape a light box so that the hard concrete pillar turns into the focus of the tender space.

The entrance hall also continues the theme of bamboo. In order to take concerted action with the bamboo ceiling, the wall is covered with wave-like bamboo which guides customers into the restaurant and shows them a new special feel.

The design of compartments stresses the difference between each other. Compartments in the 1st floor are bigger. Folded plates and cutout wall are sharing the same vocabulary from the ceiling to the wall, but each compartment is decorated with different folded plate angle and ornamental engraving patterns. The ones on the south mezzanine are smaller, using special curve of plafond and simple bamboo wall to create an interesting and elegant space. As for the suspended compartment, it is the focus of the whole space and is linked by a bridge and corridor. The translucent wall gives people, whether in the interior or in the exterior, a special experience.

We hope to create a different and united space sense and respect the local culture in the process of designing.

161

银河宾馆唐宫海鲜舫
Tang Gong Seafood Restaurant

- 建筑设计：非常建筑
- 项目主持：张永和
- 项目负责人：陈龙、何哲
- 结构咨询：郝玉范
- 面积：1 200m²

- Design Company: Atelier Feichang Jianzhu
- Designer: Zhang Yonghe
- Project Manager: Chen Long, He Zhe
- Structure Consultant: Hao Yufan
- Area: 1,200m²

上海唐宫的室内设计以"水"为关键词，与水有关的材料以及水的曲线形态成为设计的重点。由于该室内空间可以很清晰地分为特征差异很大的三个部分，于是设计师将其对应为三种有所区别的空间设计形式。曲线则作为一个统一的元素以不同的方式出现在三个空间里。

第一部分的空间是在一个夹层底下，高度很低（梁下只有2.4m左右），面积不大。在这个空间中，部分家具、墙面以及地面都以曲线的形式出现，并通过五厘板开深槽，然后在转弯处弯曲来实现。吊顶曲线是以平面图案的形式出现，材料是拉丝铝板，开槽形成发光亮线。该空间偏重于"中餐西吃"的西餐氛围，采用四人小桌，光线比较暗。

第二部分的空间较高（梁下有6m左右），面积也比较大。在这个空间中吊顶以曲线的形式出现，曲线局部上下错落形成一些发光的开口（使用白色透光膜）。四个悬挂于空中的包间被置于该曲线吊顶之上，形成了最大的几个发光开口。其曲线吊顶是通过在曲线龙骨上悬挂望砖来实现的。望砖的悬挂间距相同，但由于曲线角度的不同而形成不同的视觉密度，并在较为垂直的曲面上形成镂空透光的视觉效果。地毯和包间的玻璃隔断使用了水纹图案（后者为丝网印），也是以平面图案出现的曲线。同时，玻璃隔断又将水纹图案投影在白色透光膜之上。这个空间偏重于中式的传统聚会氛围，采用中式的大圆桌，光线相对比较亮。

第三部分的空间在室外屋顶上（属于加建），没有高度限制，面积不大。九个玻璃盒子（九个包间）由走廊连通，由竹子形成的绿化墙分隔开。每个盒子都有很好的景观质量：不仅有朝向宾馆庭院及室外的景观，而且都有一个面朝向竹子。玻璃包间的内部有一圈连续的窗帘，窗帘的颜色与包间的地毯颜色相近。九个包间有九种不同的颜色，夜晚就成为九个不同颜色的发光盒子。该空间的曲线出现于玻璃屋顶之上，是钢筋和竹片编制而成的遮阳隔栅。此空间偏重于商务特点，配置较全，每个包间都有各自的卫生间和备餐间，景观质量也比较好。

Shanghai Tang Gong Seafood restaurant uses "water" as its key word of its interior design. Materials relevant to water and curves in the shape of water become the key point of this design. The whole space can be clearly distinguished as three different parts with their own special differences, so the interior design is based on the differences and designed as three different spaces. And curves, as an united element, can be found in the three parts in different forms.

The first part is only 2.4 meters high under an interlayer with a small area. Some of the furniture, floor and wall are showed up in curve which is achieved by deep groove in 5cm-thick boards and winded at the corner. The ceiling curves made by plain view drawings whose material is pull silk aluminum board with shining lines. This space tends to make an air of western style with relatively dark light and small table of four.

The second part is higher, about 6 meters high with a larger area. In this space the suspended ceiling shows up in curve. Some of which set out up and some down thus create some opens made of white film. Four comparments that hang in the air are set above the suspended curve ceilling, making them the biggest luminous opens. The suspended ceiling is realized by suspending sheathingtile at the keel. Though the spacing between the suspended sheathingtiles are the same, it makes different visual density due to the different angles of curves and creates a visual effect of cutout at the vertical curved surface. The glass partition between carpet and compartment has splay mark pattern which shows up in curve way too, while the glass partition project the splay mark pattern onto the white film. This space attaches importance to the traditional Chinese party amosphere, using the chinese style big round table and the light is relatively bright.

The third part is on the roof (a subsidiary space). The area is big without height limited. Nine glass boxes (comparments) are united by a corridor and seperated by green wall made of bamboo. Every box has nice sightseeing facing the bamboo and the yard of the hotel and the outside scenery. With a curtain that has the same colour nearly as the carpet, each compartment has its own colour, which become 9 colourful luminous boxes. The curves made of steel and bamboo appear on the glass roof as sun-shade fence. This space has the characteristics of commercial, safe settings, independent washroom and nice scenery.

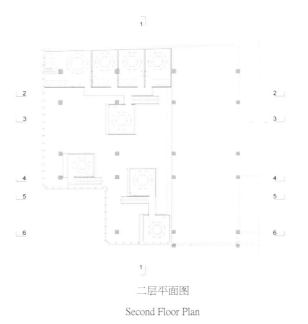

二层平面图
Second Floor Plan

一层平面图
First Floor Plan

酒醉俱乐部
Press Club

- 设计公司：BCV 建筑师事务所、Chris von Eckartsberg
- 面积：8 200m²
- Design Company: BCV Architects, Chris von Eckartsberg Design Principal
- Area: 8,200m²

本案为改建项目，希望其既能赋予创新的酒业及酒廊经营模式一个独特的城市身份，又能恰当地展现加州"葡萄酒之乡"的精神。一个极具现代质感的雕塑，既浓缩了葡萄酒之乡的休闲，又展现了"时尚"、"永恒"两大话题。本案设计借鉴Napa融产业、自然风光为一体的独特手法，让"工业化"元素的表达与"有机化"的表达齐头并进。

本案预算320万美元。巨额款项足以成就独具特色的设计。入口的零售空间，既是玄关又引领着内部空间。地下酒窖占地762m²。各个独立的品酒

室美酒飘香。8个当地世界级的酿酒厂生产的美酒给你一种"酒未入口,人已醉"的感觉。公共半隐密私人休息室,中央品酒、美食吧,私人活动室,支持空间等各功能空间应有尽有。

4模块美国产黑核桃木铺板与同样是4模块的水泥石膏板并列着,恰恰是"工业化"和"有机化"元素的表达。图案的取向、木板的表现书写着两种原本对立材料之间的微妙对话。通常被丢弃的边材在此却得到了最好的运用。延续着设计主题的统一,环保低碳也在悄悄地进行。

简约的调色板由复杂的材料勾勒。软木布、炭黑色混凝土台面、不锈钢、塑钢板在这个原木、混凝土主打的空间中恰当地找到了自己的位置。酒瓶在饰布和艺术设计品中恰如其分的运用,无声地延续和强化着空间的主题:葡萄美酒使人醉。

The goal for this venue is to create a sophisticated urban identity for this innovative wine collective and lounge business model while capturing a modern expression of

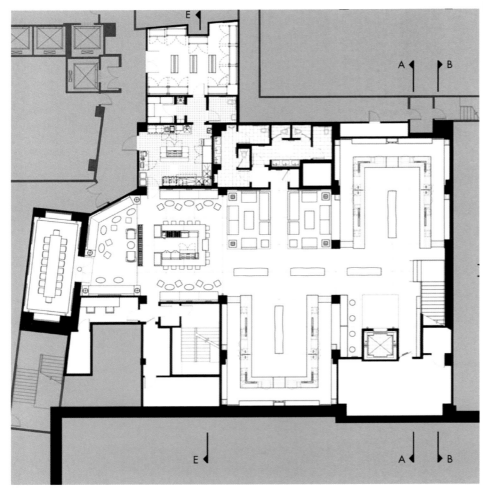

平面布置图
Layout Plan

the spirit of California "Wine Country". Within a distinctly modern sculptural expression, the architects strived to capture the casual sophistication of the wine country and create a space that is both "fashionable" and "timeless". Taking cues from Napa's unique blend of industry and natural beauty, the design celebrates the juxtaposition of the "industrial" against the "organic" as a foundation of the project expression.

With a project construction budget of $3.2 million, a sense of place has been created within a challenging urban location. With a wine retail shop and concierge at the entry, a subterranean cellar covers an area of 762m² and includes individually staffed tasting bars for 8 world-class regional wineries, public and semi-private lounge space, a central wine and food bar, a private event room and support spaces.

The "industrial" and "organic" elements are expressed in a simple pattern through 4" wide American black walnut planking juxtaposed with the imprinting of matching board formed cement plaster. Orientation shifts of the patterning from these plank expressions create the subtle dialogue between these two primary contrasting materials. The integration of normally discarded sapwood planking into the project further accentuates the pattern and conserves materials.

Other key materials compose the simple project palette. Cork fabric soffits, charcoal black concrete counter tops, and stainless and steel plate surface accents play a role in this contrast of wood and concrete. Backlit wine bottles of various colors create a striking display, the colors of which are continued in the fabrics and artwork.

Jaga推广车
Jaga Experience Truck

- 设计师：艾米
- 面积：18m²
- Designer: Ame Quinze
- Area: 18m²

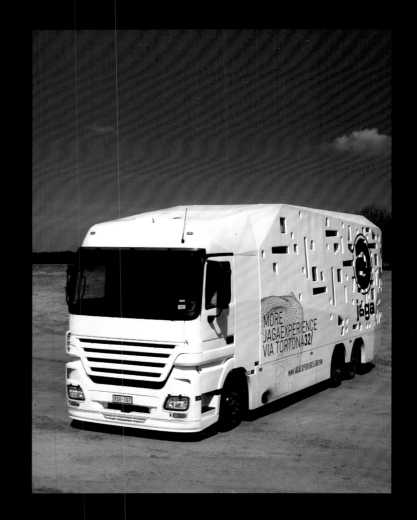

Experience Lab实为电暖器推广。该策略不重于太多推广，但一经上市，即获成功。Jaga品牌于是应运而生。本案将8t重的卡车打造为移动载体，旨在Jaga电暖器推广。设计如同螺丝壳里做道场，投影室、贵宾休息室、厨房各功能空间样样齐全。高品质的聚酯、高科技的涂层皆由手工覆盖在空间内外。90个窗口，6种不同形式，内嵌LED集成照明，多束灯光齐映星空，书写华美景象。

The instant success of the Jaga Experience Lab triggered the idea at Jaga to bring the same experience and emotions on a moving platform or medium to be exposed anywhere in the world. That led to the idea of an Experience Truck. On the basis of an 8 ton truck chassis, SAQ designed a new body holding a projection room, a VIP lounge and a kitchen. The form was handmade in high quality polyester and high tech coatings. The 90 windows in 6 different forms create together with the integrated LED-lighting a spectacular color pattern.

Brown Sugar

- 设计公司：齐物设计事业有限公司
- 设计师：甘泰来、高泉瑜
- 面积：室内198m²、室外549m²
- 主要材料：和平白石材、橡木染深色、镜面不锈钢、灰色洗石子
- Design Company: Archinexus Design
- Designer: Gan Tailai, Gao Quanyu
- Interior Area: 198m²; Exterior Area: 549m²
- Material: peace white stone, stained dark oak, mirror stainless steel, gray washed stone

Brown Sugar 台北店在设计上承袭了上海店"阶梯剧场"的概念，在开放空间内规划多种平台高度，共划分为吧台区、沙发区、阶梯包厢区、散桌区、舞台区、VIP区，基地前端架高吧台区与后端舞台区中夹着散桌区，左右两侧为架高的沙发区与阶梯式的座位。

在刻意放宽的沙发区走道与家具不贴墙的摆设方式下，动线更为自由。以不锈钢材质作为"人"字纹屏风以及立面、地台的框饰，并大量使用深色镜面及茶色玻璃，使空间中镜面反射出霓虹光彩的华丽幻影。

Brown Sugar in Taipei inherited the design concept of the ladder theater of the store in Shanghai. A variety of platform heights are planned in the open space, including the bar area, sofa area, ladder box area, scattered tables area, stage area, VIP area. Between the raised bar area in the front of the base and bar area is scattering the table area while the seats in the raised sofa area and stage area are planned on the left and on the right.

Deliberately broadening sofa area aisle and non-furnishing wall make a free circulation. The stainless steel makes the herringbone pattern screen and the façade. Floor box accessories, and extensive use of dark tinted glass mirror, reflect the glory of the gorgeous mirror.

Room 18+18 Love

- 设计公司：齐物设计事业有限公司
- 设计师：甘泰来、高泉瑜、张芃欣
- 摄影师：卢震宇
- 面积：850m²
- 主要材料：压克力、黑色镜子、黑色环氧树脂、木纹聚氯乙烯、塑铝板

- Design Company: Archinexus Design
- Designer: Gan Tailai, Gao Quanyu, Zhang Pengxin
- Photographer: Lu Zhenyu
- Area: 850m²
- Material: acrylic, black mirror, black Epoxy, PVC of wood texture, plastic-aluminium board

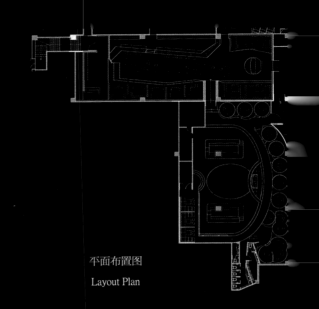

平面布置图
Layout Plan

将一系列复古式线板线条、雷射雕刻于多层次的透明压克力板上，使其成为座席间半穿透之隔屏。层层相叠融合，进而形塑出一处虚实相间、光影变幻的新英式数位花园。

Room 18 "剧场式"的空间环境架构中，透过一组双吧台的形式置入，使其和IDJ台之间形塑出一多面向、多层次舞池，同时，又衍生成新的吧台概念，即"男吧"及"女吧"。在此，人人既是观众又是明星，因为随处皆是舞台，随时皆可展现奔放的自我。

A series of vintage boards and lines and those transparent multi-lay engraved by laser, make them a translucent screen between the sea and thus create a new English style garden of changing lights and sh

In the theater-like structure, a double imbedded bar desk togeth make a multi-direction and multi-layer dance floor, which at the s new concept of bar, which are "male bar" and "female bar". Here audience and star, for every place is stage and everybody can show openly.

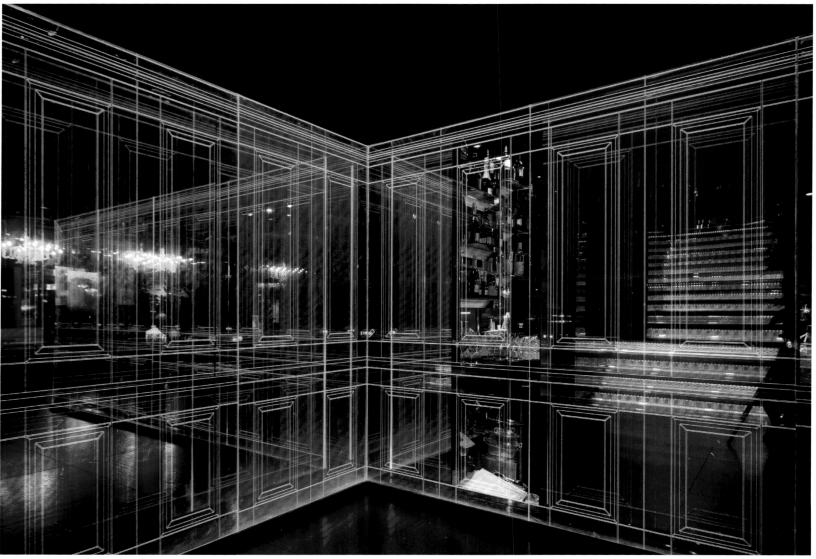

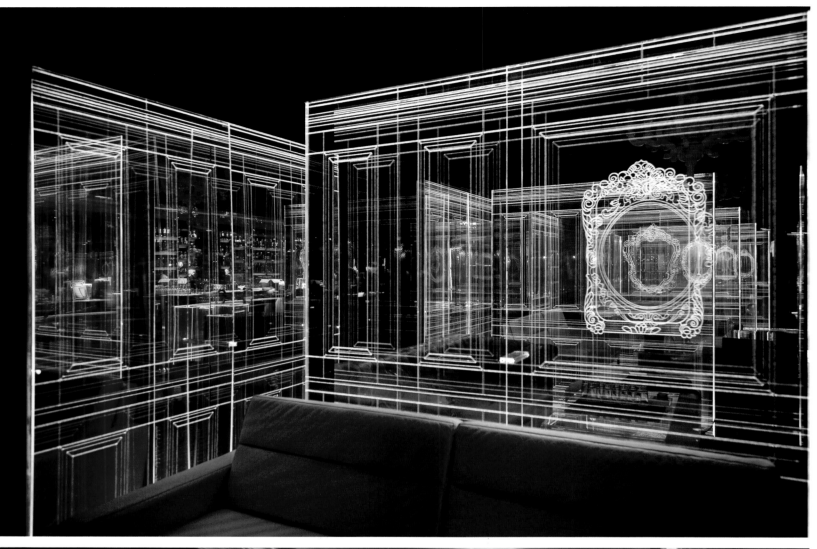

195

拉维达餐厅
La Veduta

- 设计公司：富魅公司
- 设计师：Yasumichi Morita
- 图片提供：Nacasa & Partners Inc.
- 面积：226m²
- Design Company: Glamorous Co., Ltd.
- Designer: Yasumichi Morita
- Photo Credit: Nacasa & Partners Inc.
- Area: 226m²

拉维达餐厅位于意大利，全天营业。富丽堂皇，独富艺术气质。细节考究，如精心装点的舞台，极适合上演故事。古典式的吊灯，简约高耸的天花，令人耳目一新。栩栩如生的动物，运用新型材料，绳编而生，尽显欧洲智慧。巴黎艺术大家莫扎特·格拉的作品应用其中。东方情愫，中山大介先生的手笔。棕色、米色系的内部空间，微微刮起了大阪风。后边的包房，施纳贝尔艺术书写着其中的灵魂。白天华丽、淡雅，夜晚朦胧，交相辉映。透明的窗帘使室内空间有了更多的生机与活力，阳光自然，空气新鲜。夜晚，内外灯火通明，打造一个无边际的空间，放松、舒化你的神经。怀旧与传统，轻松与微笑，尽在如斯典雅空间。

It was an elegant restaurant like an artistic palazzo. We offer all-day Italian dining at "La Veduta", the only place where you can get a brilliant sense for details as if a beautiful story begins. The classical chandeliers hanging at the simple and high ceiling will impress guests with a freshness of first sight because of the use of the innovative material organized through ropes. The beautiful animal arts made of ropes will show you European wit. It is a new attempt to use artworks of Mr. Mozart Guerra, an artist lives in Paris, for parts of interior design. The unique artwork by Mr. Daisuke Nakayama suits in the brown or beige-colored space, expressing homage to Osaka stylishly. Glamorous arts by Schnabel Effects in the private room at the rear also have great presence. Your guests will feel fresh whenever with different moods by day and night for the illumination effects. Natural sunlight and fresh air come in during the daytime through the sheer drapes, while it completely turns into moody and relaxing with the chandeliers and ambient lighting in the night time. The interior of "La Veduta" combines nostalgic and traditional designs through the glamorous filter. The elegant space with variously relaxing effects will make you smile when you step into.

〈阿诺餐饮威基基店〉
Nobu Waikiki

- 设计公司：罗克韦尔集团
- 摄影师：芭芭拉·卡夫
- Design Company: Rockwell Group
- Photographer: Barbara Kraft

非凡的感觉，尽显夏威夷主岛风情，亚洲风味美食尽在其中。本案由怀齐齐帕克酒店、阿诺餐饮、罗克韦尔集团执手联袂。不一样的体验、不一样的阿诺餐饮，尽在威基基店。

怀齐齐帕克酒店和哈利库拉尼旅游度假村两大王牌酒店长久以来期待倾情合作，如今方联手打造出这个威基基海滩精品。威基基酒店恰立于姐妹酒店怀齐齐帕克酒店对面，端庄、典雅，欢迎来自夏威夷本土及世界各地的嘉宾。阿诺餐

饮彰地方烹饪之声望,扬夏威夷之美名。

空间面积约700m²,绚丽多彩,令人眼花缭乱。本案是罗克韦尔集团的倾情奉献。世界上最抢手的国际知名建筑和设计公司的加入,实如锦上添花。餐馆、酒店、剧场等各功能空间应有尽有,一如纽约阿诺57及洛杉矶的柯达剧场之品位和风范。这里拥有夏威夷的独特质感元素,如水流、竹影、鱼网、海岛植物、石材,尽显空间之风情、幽默。

Waikiki continues to captivate the international traveler with a bounty of extraordinary new offerings such as scenic spots and Asian foods. This project is planned by Waikiki Parc Hotel, Nobu restaurant and Rockwell Group. Here you will be offered unusual experience and foods.

This collaboration with the Waikiki Parc Hotel and Resorts of Halekulani is a fulfillment of a long held dream to see Nobu come to life in Hawaii, specifically in Waikiki. Now it becomes true. Nobu Waikiki is situated on the opposite of its sister property Waikiki Parc Hotel, providing the premier location to entertain guests from Hawaii and the rest of the world. The local cuisine and Hawaii will both get a name because of this hotel.

The dazzling new 700 sq m Nobu Waikiki, full of brilliant colors, is created by the Rockwell Group with the best international architectural designers and companies involved. Functionally speaking, it can serve as hotel, restaurant and theatre, as if they are sharing the taste and style of Nobu Fifty Seven in New York and Kodak Theatre in Los Angeles. There're some unique elements such as streaming water, bamboo, fish nets, island plants and stones, showing the taste and humor of the space.

阿诺餐饮莫斯科店
Nobu Moscow

- 设计公司：罗克韦尔集团
- 摄影师：奥列格·科罗廖夫
- 面积：318m²
- Design Company: Rockwell Group
- Photographer: Oleg Korolev
- Area: 318m²

玄关

临街处一扇阿诺标志性石门引领内里空间。玻璃幕墙，竹影石门，互相映照。繁华莫斯科觅得如此一隅清冷，令人倍感清新，非常惊讶。走廊、天花、墙壁以曲线核桃木梁为导向，加之独特的发光配置，创造出海洋中生命之印象。

酒吧酒廊

天花上镶的珍珠母，彰显出海洋之曼妙，同时也为沿墙而立的众多小兔提供了恰如其分的背景。皮革、木质面板的组合挂满了墙壁，水晶地球仪悬在不同高度的空间。头顶处切边木条、染色的条纹木条，原是树皮皱纹的回忆。吧台后，雕塑的景观由瓷器、石材铸就。高贵的气质、清冷的感觉顿时蔓延开去。

主餐厅

鲨鱼皮板包裹的入口通向主餐厅。临街的窗户以灰色乌木为框架，窗楣处挂着光纤玻璃盒以感召来往的行人。天花的处理延续着玄关的主题，以曲线核桃木梁为导向，潮来潮往，生生不息。九个青铜打磨椭圆吊灯悬于木梁之间，其实是受海胆的启发。黑暗之中，暗灰色的石材地板、墙体，与发光的背光玛瑙寿司酒吧形成了鲜明的对比。

包房

第一包房墙体尤为出众，全部以核桃原木制成。另一墙上的瓷质花器雕塑定制而成，是加拿大艺术家帕斯卡尔·吉拉尔的大手笔作品。两个包房的缎质手绘壁纸，缝以樱花图案，很是耀眼。

Entrance

Guests enter the restaurant from the street through an iconic Nobu stone portal. When entering, the guests are immediately surrounded by glass walls that encase a grove with bamboos and stones in it, giving a good impression. This feature provides a fresh and surprising contrast with the bustling, urban Moscow Street. Going further, the hallway, the ceilings and walls are lined with rows of glowing curvilinear walnut beams which make it seem as if there are many sea creatures.

Bar Lounge

Mother-of-pearl covers the ceiling of the lounge, and serves as the backdrop of the many niches along the walls. The walls are covered with a combination of leather and wood panels, and crystal globes hanging at different heights throughout the space. The bar features a cutting edge wood bar on the top and a striated bar dye—reminiscent of tree bark. Behind the bar is a sculptural landscape made of porcelain and stone, giving out a sense of elegance and freshness.

Main Dining Room

Shagreen panels wrap the portal from the lounge into the main dining room. On one side of the space are windows framed by gray ebony, and topped with fibre-optical glass boxes that are visible from the street and that are perhaps meaningful for the passer-bys. Similar to the entrance hallway on the main floor, the ceiling of the main dining room is covered with curving walnut beams of different profiles, creating the illusion of a wave. These wooden beams open up to make room for nine elliptical chandeliers made of polished bronze with forms inspired from sea urchins. The glowing back lit onyx sushi bar stands in contrast to the dark grey stone floors and walls.

Private Dining Room

The walls of the first private dining room are made of walnut logs, with a custom sculpture of porcelain flowers produced by Canadian artist Pascale Girardin and mounted on one wall. Both private dining rooms feature satin hand-painted wallpaper with a stitched cherry blossom pattern.

墨西哥酒吧
Nisha Acapulco Bar Lounge

- 设计公司：帕斯卡建筑师事务所
- 摄影师：埃尔南德斯
- 面积：560m²
- Design Company: Pascal Arquitectos
- Photographer: Sófocles Hernández
- Area: 560m²

平面布置图
Layout Plan

　　墨西哥酒吧坐落于墨西哥阿卡普尔科，作为娱乐场所，全力打造感官体验。建筑量体、人居活动和音乐之声和谐共存。

　　玄关模拟船舱样式。五个高清显示屏如椭圆开窗框景。仰望处，如浮云行走；俯看时，如深海潜艇。天之蓝，海之深，尽藏其间，彰显对比。

　　玄关旁设休息空间。室内布置古色古香，装点着几组小型的沙发。踱步即至的室外吸烟区清晰可见。一端悄悄地立着几扇隔屏，延续着玄关的框景主题，无形中使家居客厅的气氛蔓延开去。另一端，15m长的吧台后，等长的高清晰度视频高达3m。LED照明色彩明亮，让几何设计的内里空间瞬间有了生命的质感。

　　洗手间为玻璃空间。女士区红白相间，中心区设一株挺拔的巨型棕榈，并安放着长长的小凳，尽显女性娇柔、纯洁气质。男士区为黑、蓝二色世界，三角形中央有装饰性绿植，独富异国情调。

　　音乐在空间叮咚流淌，为图像、视频静静地伴奏。惬意的体验、静谧的空间就在这里。

This lounge bar, located in Acapulco, is an entertainment place dedicated to the senses. Expressed through the architecture, the people, the music and the images, an alternative and virtual atmosphere is created.

Access to the entrance that simulates the interior of a ship lined with wood, where five high definition screens were framed as a view of oval windows that may show either a clouded sky travelling at high speed or bottom of the sea images transforming it into a submarine.

A dark foyer leads to a wooden lounge decorated with small groups of sofas that have a clear view of an outside smoking area. At one end there are several screens framed as pictures in a living room and at the other end, a 15 meter long bar counter and further up, at the back, video screen of the same size of 3 meters high. The under

counter bars display a geometric design and are lit up by changing color LEDs.

Restrooms are of glass place; the women's area is the area that is co-ordinated with red and white colors and equiped with a bench and a huge palm tree at the center, while the men's section is in a black and blue place with a central triangular decorative plants with exotic plants.

The music is on accompanying with images and video. A comfortable experience is for you in the space.

桑普酒吧
Thumper

- 设计公司：新加坡王氏私人有限公司
- Design Company: ONG & ONG Pte Ltd

本案灵感源于英语单词"Thumper"与兔子的关系，在英语中它往往是兔子的代名词。鉴于此，设计者希望本案能够给人一种"静如处子，动如脱兔"的感觉。

进入俱乐部，你会看到休息室如穿线般串连起所有主要功能空间。一如其他酒吧，酒吧空间分为主酒吧区、舞池、休息室和私人休息室，但设计让你进入酒吧时有如玉兔入巢穴的现场感。空间体现出立体螺旋形的设计特点，愉悦着你的感官，激起你入内探奇的欲望。休息室如螺旋般推进，既好玩又能拉近你和酒吧空间的距离。

Taking inspiration from that analogy, the design concept of the club was to create an environment where the visitor assumes the characteristics of a rabbit.

Entering the club, the visitor walks through the transitional lounge that links all the main areas together. Serving as a primary spine, the visitors are channelled into the three main sections of the club: the main bar/dance floor, lounge and the private lounge. In keeping with the name and concept, design of the transitional lounge was inspired by a rabbit's journey through tunnels to reach its burrow. A three-dimensional spiral form flows subconsciously, enticing visitors to venture deeper into the club. The spiral treatment of the transitional lounge is further reflected in the columns thereby providing a playful, dark and intimate atmosphere to the club.

平面布置图
Layout Plan

赤鬼炙烧牛排崇德店
Oni Boy Steak Chongde Shop

- 设计公司：活设计国际创意有限公司
- 设计师：张顺云
- 面积：一楼室内300m²；二楼室内170m²；户外空间271m²
- Design Company: Ho Design
- Designer: Zhang Shunyun
- First Floor Interior Area: 300m²; Second Floor Interior Area: 170m²; Outdoor Area: 271m²

日式牛排特别强调炙烧火烤的做法，这在中国台湾还是全新的概念，单店月售四万客的数字说明它拥有广大的市场。

在日本战国时期，"鬼"是许多骁勇善战的将军的别号。店名与日本品牌"鬼洗"牛仔裤的"鬼"异曲同工，都具有如鬼神力之气势、叛逆反骨之创新性格。日式风格不再只是竹子、锈铁、清水模，而是贴切地传达了平民大众对于平价消费的更高期待。

本案设计将牛排幻化成日本武士，顽强地与炙热焰火搏斗。外观以充满力量的红黑铠甲象征武士，火红的不规则光带如同烈焰缠身，以戏剧性手法造成强烈的视觉冲击，吸引街头目光。

用餐空间延续外观的不规则语汇，以木纹、金色及灰阶色调，透出另类日式的用餐氛围。

层叠错落的天花造型，犹如武士格斗的刀光剑影所划开的裂缝，当然这也有聚气排烟的机能考量。

店内的柱体与服务柜台都采用不规则的几何块面包装，晶亮的黑色材质与白色的天花形成鲜明的对比，赋予空间桀骜不拘的个性。当飘着浓香、嗞嗞作响的红色牛排端上来时，将迅速催化空间的热情，提升对美食的欲望。

就餐区的卡座排位非常理性和有序，空间得到了最大化的有效利用。理性的排位也有利于客人迅速找到位置就座，这对于像赤鬼炙烧牛排这样常常人潮爆满的店来说尤其重要。理性的布局并没有令设计师停顿创意的构想。卡座之间一排高低起伏的长钉装饰与赤鬼武士桀骜勇猛的气质相呼应，让空间有了音乐般的律动。从天花上悬吊而下的立体白色吊灯，修长的线条，显示出细节的精致。

同样是赤鬼炙烧牛排店，设计师在公益店与崇德店之间用不同的设计语言描绘了两种迥异的空间型格，公益店是火上浇油式的外放热情，崇德店是水包火式的内在沸腾，不一样型格，却同样受到食客的热烈追随。正如"活设计"的公司名称一样，度身定制，灵活设计，让每一个空间都拥有个性的光辉。

The Japanese steak, which emphasizes flame grilling, is a brand new concept in Taiwan, China; sales figures of 40,000 servings per month speak for its high market position.

During the Japanese Warring States Period, "Oni Boy" was an alternative name of brave and battle generals. The name of the restaurant is similar to a Japanese jeans brand, which has the strength and vigor of "Oni Boy", as well as a rebellious and innovative personality. The Japanese style is no longer an interpretation of bamboo, rusty metal and precast concrete and now conveys the public's higher expectation for low prices.

The steak is transformed into a Japanese Samurai Warrior who stubbornly battles the red hot flame. The exterior utilizes powerful black and red armor to symbolize the warrior, while the scarlet, irregular light belt engulfs the body like flames. The dramatic approach creates a powerful impact and attracts the attention of people.

The vocabulary of irregularity continues in the dining space, and an alternative Japanese dining ambiance is created through the application of wood grains, gold and grey tone colors.

The overlapping ceiling design is akin to a fissure created by the swords of fighting Samurai; in fact, it has functions of fumes accumulating and giving out.

The columns and the service counter are made in the irregular geometrical shape. Clear and brilliant black materials are in the contrast with the white ceiling, creating a tameless and open personality for the space. When the aroma is filling with the space and the buzzing red steaks are on the way to your table, it soon will increase your passion and desire for the food.

The arrangement of the seats in the dining area is of great reason and order, making the space in the best use. Because of the arrangement, clients can find suitable tables quickly, which is far more important for this project. Although the arrangement should be reasonable, it's still full of innovative ideas. The decorative spikes in a high and low row are sharp contrast with the braveness of the warriors, making the space bursting with the rhythm of music. The white chandeliers hanging below the ceiling and their long lines show the delicacy of the details.

Even though the same such kind of steak restaurant, there're still some differences between Gongyi shop and Chongde shop. The former has a style of openness and passion whereas the latter has a style of inside enthusiasm. Both of them of are liked. As the company name "Ho Design" ("Ho" means flexible) suggests, it's custom-made and flexible, as if every part of its space has its own personality.

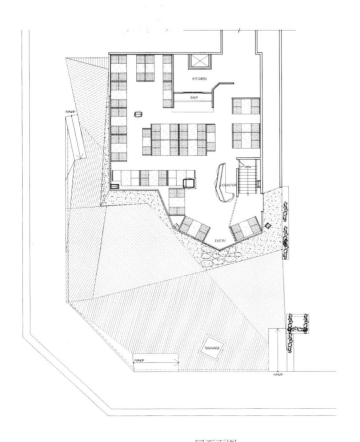

一层平面图

First Floor Plan

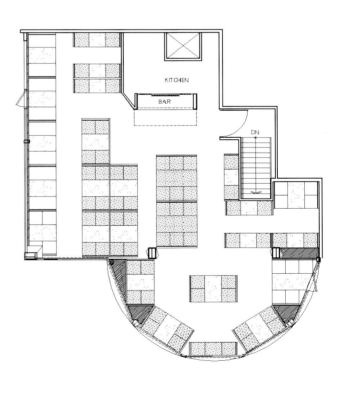

二层平面图

Second Floor Plan

赤鬼炙烧牛排公益店
Oni Boy Steak Gongyi Shop

- 设计公司：活设计国际创意有限公司
- 设计师：张顺云
- 面积：574.1m²
- Design Company: Ho Design
- Designer: Zhang Shunyun
- Area: 574.1m²

如今，人们在就餐消费之余，同时也在品味生活，注重用餐环境带来新的五感体验的"食尚"乐趣。当餐馆设计不再只是闭门造车和照本宣科时，突破常规的大胆创意是时代转折的起点。试想在就餐环境中，举目四望，线条与方块相互交织，相映成趣，餐厅的镜面反射出五彩缤纷的灯光，营造出梦幻气氛，华丽的设计和空间布局成就了磅礴气势，这样美妙的就餐视觉体验将引领新的"食尚"潮流。

赤鬼炙烧牛排店以鬼为名，外观以纯黑金属麒麟板不规则倾斜塑造出刚硬风格，地坪运用南方松表现出不规则的层次感，呈现出地面的裂纹，不规则倾斜的黑色量体以钢索箝制，仿若即将倾倒的招牌，加上路旁斜格纹意象的红色立体压克力灯箱，以上种种戏剧性的手法突显了就餐的场景，造成了强烈的视觉冲击效果，成功达到吸引路边行人目光的初衷。

"鬼"的称号在日文中有赞美之意，意谓有英明勇武之气势，见其人

如见鬼一样令人畏惧三分。张顺云表示，过去在日本许多战国名将都有鬼的称号，如德川四天王之一、德川三杰之一的井伊直政，人称"赤鬼"，正是借"鬼"的称号表示对勇者敬意的一个例子。于是设计师依循业主设定的日式料理风味牛排专卖店概念，将"鬼"在日本文化语境里的象征意涵融入其间，以创新颠覆的处理手法创造出磅礴的气势和视觉上的强烈冲击感。

走进室内，火与日式刺青的意象被大量使用，强调被火纹身带有浓郁日式风味的牛排。天花板使用的工作灯并排陈列，整齐划一，表现出大气华丽之感。整个虚幻红色量体和漂浮空中的垂吊红色链条形成点矩阵状排列，放大了空间视觉效果，宛如一场缤纷的烟花红雨，既实现了遮蔽天花喷黑结构的功能，又让人在视觉上得到美的享受。

架高区定制的大型火把灯，如路灯般整齐排列，虽然火把一向给人粗犷的印象，但此处的火把灯如卫士一般给人秩序和安全之感。空间尽头则运用火红、金色相交错的马赛克壁纸，呼应主题，张力十足。整个空间天雷勾动地火的黑红相遇，在设计好的情节里诱发空间能量的沸点，以时尚角度重新诠释平价牛排馆的新形态。张顺云抛开中规中矩的表现原则，借助空间存在的形式与物件之间的关联，大胆呈现出极具五感体验的氛围，为就餐者带来一场视觉盛宴。

空间自身可由人任意设计和布置，设计师可以通过把不同的设计元素融入空间设计中，带来让观众惊艳的观感。在没有任何画面的初始，空间就是一只空盒子，但设计师能够凭借对内部物件的铺陈排列，通过方向上的调整及加诸其内的技术性巧妙构思，创造出前所未有的空间体验。对张顺云而言，空间是一个舞台，只要设计师懂得运用设计元素，以各种各样的方式表现出天马行空的创意，就能吸引台下的目光，迸射出动态与静态间碰撞产生的火花，并和人们产生共鸣。

Lines, blocks and mirrors are a good reflection of breathtaking momentum. Appearance and its inherent values increasingly have no way of avoiding cuisine fun exacted on five senses. When design is no longer subject to process made by only one person without exchange with others, and is free of repeating what book says, roaring expression without other regard is the bold turning point of times.

Its façade is clad with black metal unicorn plates irregularly tilting; the southern pine flooring looks like a random perfection; the sloping building is of steel cabled with slanting signs; the entry is impressive with a red image of oblique acrylic light box, on which the shop name of Oni Boy Steak is dramatic to make an impact on passers-by, because it is eye-catching and visually effective to be kept in mind.

Historically, "Oni Boy" in Japanese has been praised with its good intentions, which means a wise chivalrous momentum. According to Zhang Shunyun, many famous generals in Warring States are honored as "Oni boys", such as one of four kings of Tokugawa. Such a cultural connotation makes a solid basis for the design. Images of "Oni Boy" in Japanese culture are thus implanted, creating an imposing spatial impact with the approaches of innovation and subversion.

The interior is full of images of fire and Japanese tattoos, which are a strong emphasis on Japanese-style steak, while ceiling lamps in row bring out a gorgeous sense. Red chains in a dot matrix are actually an illusion of red building volume, whose effects, like that of colorful fireworks, are both aesthetic and practical to hide smoky ceiling structure.

Custom torch lights in the elevated area are soft just like street lamps. At the end is the application of mosaic wallpaper, red and gold, that is filled with tension and echoes the theme. When black meets red, boiling point of energy gets immediately induced. The fashion is incredible to be a new interpretation of a steak restaurant that's not expensive. While law-abiding principle is abandoned, the relationship between space and furnishings is successful to present an ultimate atmosphere five senses can experience.

The inborn flexibility in the space creates an unprecedented feeling through various design elements. Space, like an empty box, makes a new experience, when displays and furnishings seem to grow out of adjusting direction and technical ingenuity. "Space is a stage and can be attractive to convey a dynamic and static spark when striking a resonance. When design elements are used in various forms of great tension, it can attract the audience's attention." says Zhang Shunyun.

前门M餐厅
Capital M

- 设计师：黛布拉·罗杰
- Designer: Debra Little, Roger Hackworth

拥有逾600年历史的前门大街，是京城最著名、最繁华的商业中心。前门M餐厅位置得天独厚，位于新近改造的前门步行街中，与正阳门箭楼仅一街之隔。三层加露台的结构，在避免嘈杂的同时可以令你无障碍地欣赏广场附近的街景。站在餐厅的露台上，你会发现设计师选择视角之巧妙，仿佛早已预计好了广场最精华的内容：露台将箭楼、正阳门、天安门的景致尽收眼底。白天视野开阔，纵览万里无云的天空；傍晚华灯璀璨，瑰丽盛美的风景如同一场古都派对。

对于餐厅室内的设计，两位已在亚洲工作8年的设计师Debra Little与Roger Hackworth更显游刃有余。他们将中西文化的感悟融于一体，在前门M餐厅营造出浓厚而典雅的就餐氛围。由知名艺术家Michael Cartwright创作的大型壁画《Journey》贯穿整个空间。在空间的转折切割中，以印象派的手法诠释出一条河流四季的变幻。带有几何图案的黑白地板传递着视觉上的理性之美。天花板处精致的银色团花纹加上大面积的镜面，使空间更显开阔，并融入了中式印象。在中心大厅，米字格栅的天花与地面的波普花纹相呼应，仿佛进行着中西文化的对话。

这里的家具配饰也极为精致，沙发桌椅中麂皮、天鹅绒与暗色实木的大量应用同样突出了华丽的宫廷气氛；造型简洁却更显收敛，沉稳的现代风格洋溢其间。

The history of more than 600 years endows Qianmen Avenue a status of the most famous and the most bustling commercial center. Capital M restaurant, a newly-transformed project, is uniquely positioned in a setting in the new pedestrian street across the watchtower of Zhengyang Gate. Its three stories plus a terrace are accessible to the view of the street but without noise. The terrace is remarkable to be a good position to take a panoramic view that stretches to Zhengyang Gate and Tiananmen Square. During the day time, it presents a cloudless sky; in the evening it goes with splendid lanterns. Nothing, modern or ancient of the capital, can escape from the sight within.

Frankly speaking, nothing about space design can throw challenges to Debra Little and Roger Hackworth. The working experience of 8 years in Asia gives them a good perception of integrating the Eastern and the Western culture. Dining atmosphere in M Restaurant is strong and elegant, where large mural *Journey* painted by Michael Cartwright, a well-known artist, were hung in the whole space. The approach of impressionist shows a changing seasonal river. The black and white floor visually displays a sensible Chinese beauty. The ceiling is embellished with delicate silver floral patterns as well as large area of mirror. The grating ceiling of the central hall echoes with the bop pattern on the ground, and a dialogue seems to be on the way between Chinese and Western culture.

Furniture accessories continue the delicacy: chamois of tables and chairs, velvet, and dark woods are employed in a large amount. All confide in an ornate palace atmosphere. Modeling, all the same, is simple, low-key, steady, sedate but still modern.

牛排城四望亭店
Steak City Siwangting Shop

- 设计公司：上瑞元筑设计制作有限公司
- 设计师：孙黎明、胡红波
- 摄影师：戴俊峰
- 面积：800m²
- 主要材料：仿古叠石、地砖、中花白大理石、橡木板、叠纹墙纸
- Design Company: Shangrui Yuanzhu Design Co., Ltd.
- Designer: Sun Liming, Hu Hongbo
- Photographer: Dai Junfeng
- Area: 800m²
- Material: antique stone, floor tile, marble, oak board, moire wallpaper

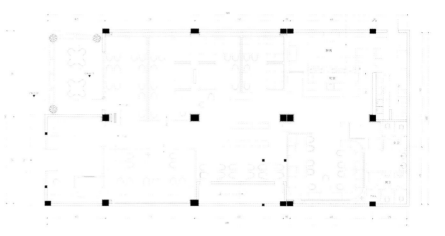

平面布置图
Layout Plan

整个空间最突出的就是十字拱顶。拱顶体量巨大，从柱面延伸到顶面，既具有造型的美感，又强调了空间的仪式感。拱顶一般是西方建筑的主要特征，这不但体现了建筑的庄重，而且影响着用餐者，使他们用餐时的尊贵感被提升，还表述出这个餐饮空间的高雅和与众不同。

顶部灯具的形态经过精心考究。大面积垂直向下的灯具形态，配合着拱顶，打破了空间的沉闷。地面黑白格子拼花，有着强烈的视觉冲击效果。精心挑选的陈设，是整个空间气质的点睛之笔，使用餐者的视觉和味觉得到极大的满足。

The most prominent character in this space is cross arch vault, whose volume is pretty big, extending from the cylinder surface to the top ceiling. The very aesthetic shape emphasizes the sense of ritual space. The vault is generally the main features of Western architecture, which can reflect the architecture solemnness, and make effect on the diners who will have a sense of their dignity. All of those can express the elegance and difference of the restaurant space.

The forms of lamps at the top have been taken seriously; the many vertical downward lights, together with the vault, break the boring space. Black and white plaid parquet floor has very strong visual impact. Carefully selected furnishings are the temperament of the entire space, giving the diner a great satisfaction of visual and taste experience.

米斯特比萨店
Mr. Pizza

- 设计师：Yeonhee Keum、宛艳玲、孟庆礼、和维
- 摄影师：贾方
- 面积：721m²
- Designer: Yeonhee Keum, Wan Yanling, Meng Qingli, He Wei
- Photographer: Jia Fang
- Area: 721m²

米斯特比萨是在韩国位居龙头位置的比萨连锁品牌。自2000年进入中国市场，以其精良的手工制作、新颖的店面设计逐渐赢得消费者的认可。该品牌主要针对女性顾客群，主推清淡爽口系列口味比萨，以100%手工制作、100%现场定做为特色，更以"love for women——为特别的她"作为自己的品牌诉求。而米斯特餐厅的装修风格也是为了呈现意大利手工比萨餐厅的氛围，同时突出舒适自然的气氛和趣味干练的现代风格。

呼和浩特的这间米斯特餐厅位于中山路海亮广场的六楼。这里的店面设计是总部派韩国设计师专门设计的，与国内很多其他比萨店相比，无论在整体色调上，还是在物品摆放上，都有种与众不同的感觉。现代的设计语言，缤纷如糖果般的色彩装饰，墙上的彩绘，灯光剔透的卡座包厢，都展现了想象、活力、时尚的年轻气质。

一进入餐厅，迎宾处的红色背景墙，还有不规则流线型的木座椅，都让人觉得新鲜好玩，充满了想象空间。大厅的墙壁上，绚丽的都市彩绘供人在等待比萨上桌之前欣赏和消遣，而卡座区的半开放包厢里每一个的色彩和墙绘都不一样，在灯光的烘托下犹如一个个光彩四射的糖果色童话小王国。餐厅内部的设计还融入了蒙古族元素，有两处卡座是以蒙古包形式体现，这在众多座位里显得独树一帜。

Mr. Pizza is a leading pizza brand in Korea. It has won customers' hearts through its hand-made way and special interior design since its entering Chinese market in 2000. Taking women as its potential clients, 100% products made by hand and made for a while after the customer orders have become its biggest specialties and "love for women—for special her" is its philosophy. The interior possess the atmosphere of traditional Italian restaurant and the comfortable and natural air of modern style.

Located on the 6th floor of Hai Liang Plaza, Zhongshan Road, this restaurant is designed by the Korean interior designer appointed by the headquarters of Mr. Pizza, thus the design, the whole hue and the layout of goods are all different from other branches. Modern design, colorful decoration, colorful patterns on the wall and bright compartments all show the air of imagination, fashion and vigor.

On entering, the red background wall and irregular wood chairs will make you feel funny and interesting. The colorful patterns on the lounge wall are the enjoyment and pastime for customers waiting to be served. Every half-open compartment has different color and wall pattern, which under the light like a small candy-color kingdom in a fairy tale.

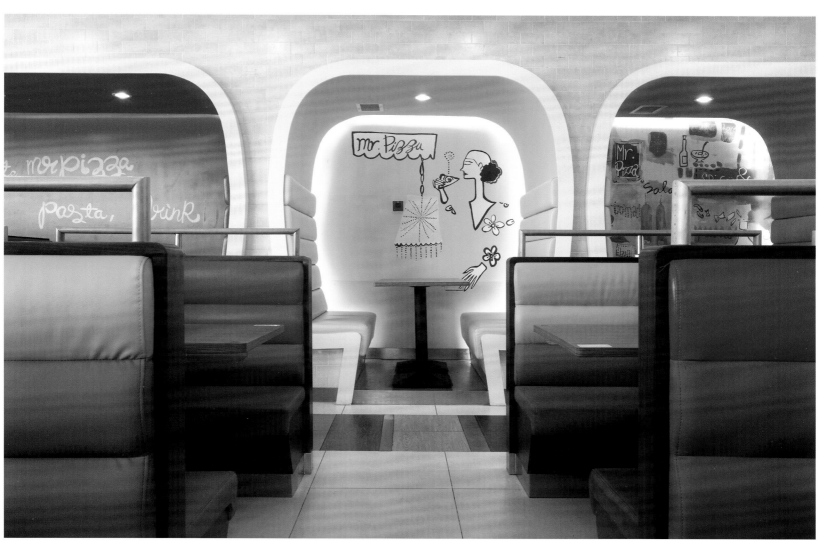
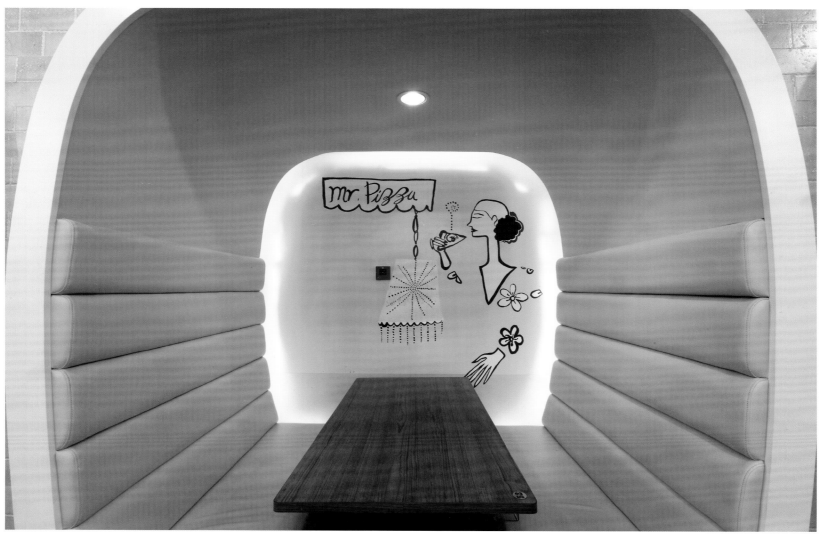

意粉屋
The Spaghetti House

- 设计公司：何宗宪设计有限公司
- 设计师：何宗宪
- 参与设计师：Momoko Lai、黎凯婷、梁姨婷
- 摄影师：Graham Uden, Ray Lau
- 面积：187m²
- 主要材料：油漆、清玻璃、清镜、红色夹胶玻璃、药水砂图案玻璃、可丽耐、布料、木纹防火胶板、亚克力胶、马赛克、墙纸、地砖
- Design Company: Joey Ho Design Ltd.
- Designer: Joey Ho
- Assistant Designer: Momoko Lai, Li Kaiting, Liang Shuting
- Photographer: Graham Uden, Ray Lau
- Area: 187m²
- Material: paint, clear glass, clear mirror, red glass, acid etched pattern glass, corian, fabric, wood grain formica, acrylic glue, mosaic, wallpaper, floor tile

为协助意粉屋的品牌重塑，本案设计旨在创造一个既简单又细致的意大利美食环境，为意粉屋创造一种全新的用餐体验。总体设计采用现代摩登的手法去演绎一种崭新的用餐格调。餐厅以红、白两种简单而富鲜明对比的颜色作背景，把整个用餐场所转化成一个充满欢乐气氛的空间。

餐厅被分为数个不同区域，以满足顾客的不同需要。入口处附近的空间被划分为私人区域，以酒廊式箱座设计，给顾客提供一个亲切舒适、可以与友人喁喁细语、开怀谈心的私密空间。

这种愉快的格调正好吻合了典型的意大利人喜欢庆祝、热爱自然、爱好社交的生活模式，并透过空间设计把这种特有的生活态度更为活化。设计师以意大利人的传统聚集点"中央广场"（"piazza"）作为创作灵感，把餐厅的中央设计成主要的"用餐广场"，并以简洁而富魅力的图案合成的家具及空间饰品，在尊贵及舒适合奏的混合音乐下，演绎出"随意聚会"的空间感。

在仿效意大利人的悠闲生活及享受自己的生活态度时，这个现代的摩登空间也同时加入了一系列的自然元素，以加强意大利的生活韵味。餐厅的墙身及镜子均印有花卉图案及绘图，包围着整个室内空间，借以添加放松神经、令人心旷神怡的精华，来触动用餐者的感官。

空间里的平面绘图不仅被用作大自然的映象，更以当代的艺术手法显示线条与意粉的视觉联系。餐厅的墙纸及墙面凹凸形的装饰板条设计，以垂直线条隐约呈现出意大利粉细长的形象；卡座上方的天花上则以缠绕弯曲的线条滑行交叠，以一幅由曲线绘制的画品，反向地映照出用餐时的愉快情境。餐厅中央建有一道特色墙壁。壁面以幼细的笔触线条绘成一幅大型壁画，横跨整个大厅，仿佛要把整个空间包围，并使之变成一个充满意大利粉风味的世界。意粉屋的传统卡通画及彩花吊灯仍然保留，并被视为餐厅的独有特色。意粉屋餐厅于空间上别具创意的元素混合，使整个空间不仅为用餐者带来一种充满欢笑的用餐体验，还带来一段前所未有的感官旅程，同时使现代化的生活质量获得更深体现。

A simple and delicate Italian cuisine environment is nothing but a brand rebuilding of a new dining experience in the Spaghetti House. Such a realization eventually turns to be a modern approach. Contrast between red and white of the backdrop turns the whole fields into a dining atmosphere full of joy.

The division of several areas meets customers' needs. The space near the entrance serves as a bar lounge that is cozy, intimate and comfortable for friends to chat with privacy highly respected.

This typically delightful Italian style meets up with Italian tendency to celebrate, to love nature, and to enjoy social life. This unique attitude towards life becomes more active through design. The traditional Italian gathering point of "piazza" is taken as inspiration, and the central is made into a dining square that houses furniture and ornaments in concise and attractive patterns. Music mixing with comfort and distinction makes a sense of space of "free parties".

The desire to follow the attitude of Italian life and their way of enjoying life, adds a series of natural elements into the modern space to enhance the flavor of Italian life. Walls and mirrors are printed with floral patterns and graphics, relaxing and refreshing, exerting an impact on sensual touch for diner.

The graphic design not only is used as a natural image, but with a contemporary artistic method displays the visual contact between lines and spaghettis. Wallpaper and wall in concave-convex plate form, vaguely shows the long and thin spaghetti through vertical lines. Deck ceiling reflects a happy dinner with winding curved lines. In the center stands a wall, on which lines are fine but imposing to make a large mural, spreading across the hall and thus making here a pure spaghetti world. The traditional cartoons and hanging lamps usually popular in a spaghetti house are retained, which is considered a unique characteristic of this place. Additionally, the creative mixture of elements presents a dining experience filling with laughter, and embodies requirements for modern quality of life, bringing an unprecedented organ journey.

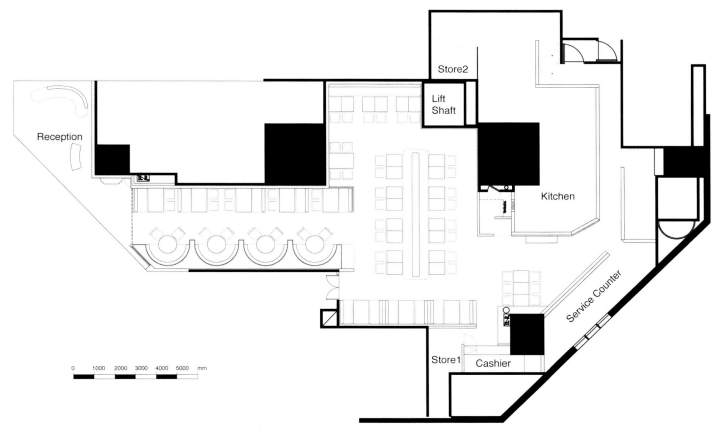

平面布置图
Layout Plan

怡亨酒店明园餐厅
Yi Heng Ming Garden Restaurant

- 设计公司：Thomas Schoos Design
- 图片提供：台湾怡亨酒店
- Design Company: Thomas Schoos Design
- Photos: Yi Heng Hotel Taiwan

怡亨酒店是台北第一家现代艺术精品酒店。全新推出的粤菜餐厅明园，同酒店本身的艺术风格如出一辙。明园的设计同样也走艺术路线。如果说大堂电梯口的达利1980年的两尊雕塑作品《Alma Del Quijote》和《Mercurio》是酒店的镇店之宝，那么Andy Warhol 1984年的作品《The Queen》则是明园餐厅最贵重的艺术品之一。而Andy Warhol还只是一个强悍的引子，来自世界各地的雕塑、装饰、油画等现代艺术作品济济一堂，令整个餐厅更像一个艺廊，走进来把作品看一圈，大概连吃饭这件正事也会忘记了。

餐厅位于酒店的地下层，一个巨大的旋转楼梯引人步入艺术境界，也令藏于地下的餐厅并不显得逼仄。为明园操刀设计的是洛杉矶设计公司Thomas Schoos Design，这个擅长于餐厅和名人居所设计的公司将明园的六个包房环绕旋转梯所形成的采光空间而布置，每个房间里都有巨幅的当代绘画艺术作品。中间的公共餐区的四张餐桌设计出四种不同的主题，"骏马之桌"、"亚洲之桌"、"绅士之桌"和"当代之桌"分别以相应的艺术品陈列来作区分。古董家具和现代设计家居、灯具、餐具穿插其中，使整个空间具有boutique shop的奇妙风格。

除了琳琅满目的古董及艺术品外，餐厅的墙身饰以牛皮与斑马皮，其中一组餐桌及座椅更以骆驼骨雕刻而成。包括天花板上的古老动物标本，全

部搜罗自台北及香港的古董店。值得一提的是,这些饰物都是酒店老板黃健华的私藏,餐厅中不可忽视的还包括顶级陈年葡萄酒和名贵的茅台酒。

Yi Heng Hotel is Taipei's first modern art boutique hotel. Ming Garden is a traditional Cantonese restaurant and the design style is exactly the same to the hotel. Ming Garden is also taking the art route. If you say the sculptures *Alma Del Quijote* and *Mercurio* of Dali in 1980 which are placed in the lobby are treasures of the most valuable works in the hotel, then the works of Andy Warhol *The Queen* 1984 is one of the most precious works of art. The Andy Warhol is just a powerful primer. Sculptures, decorations, paintings and other modern works of art from the world all gather here, making the whole restaurant more like a gallery. Maybe you will forget to have dinner when you

are appreciating the famous art works.

The hotel restaurant is located in the basement; a huge spiral staircase leads people to the palace of art, which makes the underground restaurant seem not cramped. The design company Thomas Schoos Design from Los Angeles is specializes in restaurant and celebrity home design. Six rooms around the well-lit space formed by the rotating staircase all have huge contemporary painting works of art. Four tables in the middle of public dining area are designed to four different themes, "Horse Table", "Asian Table", "Gentleman Table" and "Contemporary Table" respectively, distinguishing works of art on display. And the antique furniture and modern design furniture, lamps, tableware mix up with each other, making the whole space become a boutique shop.

In addition to an array of antiques and works of art, the restaurant walls are decorated with leather and horsehide. One set of table and chairs is carved by camel bone; the ceiling is decorated with the old animal specimens, and all of the decorations above are collected from Taipei and Hong Kong antique shop. They are all the possession of the hotel owner Huang Jianhua. And the things that can not be ignored are the vintage wine and expensive top Maotai.

中意汇意式餐厅
River Club Restaurant

伴随中意汇意式餐厅在北京的盛大开业,米其林二星餐厅Piccolo Lago的意式高级料理风尚,如今从皮埃蒙特的梅尔戈佐湖畔来到了北京麦子店的亮马河畔,毗邻使馆区域。中意汇内部面积850m²,可容纳91个座位,另外还有一个600m²的露台供室外用餐和举办各类活动。

这是Piccolo Lago首次向海外扩张。Piccolo Lago是一个乡村式的家族餐厅,在明星创始人兼主厨Marco Sacco的带领下,Piccolo Lago享誉国际,跻身意大利十大顶级餐厅之列。

"两种文明汇合、联结两个世界,是这个餐厅项目的根基。"他介绍道。在他看来,此次烹饪上的东西合璧,象征着两个有着数千年文化历史国度之间的纽带——700年前马可波罗历险到中国,将中意两国首次联系在一起。

中意汇的青蓝色标志及其内部装饰都体现出"水"元素,让人仿佛置身于水族馆,向步入Marco Sacco美食殿堂的人们,传递着一种具有象征意义、甚至神圣的尊贵感。

上行至餐厅的途中,客人们会经过一个蔚为壮观的酒窖。酒窖中藏有4 000瓶精选佳酿,墙上的世界地图则标明了各自的产地。天花板上的灯光犹如星星般闪烁,让人联想起在同一苍穹下,欧洲文明与亚洲文明相依共存,源远流长数千年。酒窖提供私人储藏服务,享此尊权的客人可将其私人贵重酒品寄放在玻璃"保险箱"内。

最大最尊贵的两间私人餐室,视野畅通无阻,有着独立会客厅的轻松装饰风格,坐拥亮马河及北京城的全景式景观。

主餐室的一侧是餐厅入口,拥有一个舞台般升起的开放式厨房,向所有人展示来自Piccolo Lago的传统烹饪理念与创新烹饪理念的奇妙融合。客人们在赞叹那些给意式烹饪带来革新的最前沿技术及新一代炊具的同时,也能欣赏到大厨们的"古法"技艺。

餐室的装饰布置也借由古典和现代主义理念,体现出两个世界的融合,从典雅的路易十六世风格扶手椅,到主用餐区欧洲极简主义风格的梦幻白和石材,以及私人餐室的马赛克墙面和木制镶板,都体现这一点。大空间的私人餐室中铺着红杉木地板,小凹室中摆放着舒适、夹棉、具有东方情调的扶手椅。

As the opening of River Club Restaurant in Beijing, Michelin Award's Italian Restaurant Piccolo Lago now comes to Landmark River from Piedmont Lake, which is adjacent to the embassy area. The internal area of 850 square meters accommodates 91 seats, and a terrace of 600 square meters for outdoor dining can organize various activities.

It is the first time that Piccolo Lago expands to overseas. Piccolo Lago is a village-style family restaurant, with the leader, the chef Marco Sacco, and it wins a great international reputation, and ranks among the top ten of the Italian restaurants.

"Confluence of two civilizations, connection of the two worlds, this is the foundation of this restaurant." In his view, the cooking in the East meets the taste of the West. It is a symbol of both cultural and historical ties between the countries. 700 years ago, the Adventures of Marco Polo to China created the connection between

China and Italy for the first time.

Italian blue logo and interior design reflects the "water" element, and people feel like being at the aquarium. It conveys a symbolic meaning, and a noble and sacred sense to those visiting it.

On the way up to the restaurant, guests will go through a spectacular wine cellar, which is of the possession of 4,000 bottles of best wines with their origin marked on the map of the world on the wall. The lights on the ceiling shine as stars, reminding us of that the European civilization and Asian civilization coexist in a long history of thousands of years in one world. Guests can enjoy private storage service to keep their valuable wines in glassy "safe".

The two largest and most prestigious private dining room with separate sitting room decorated in easy style has unobstructed view of panoramic landscape of Beijing.

Restaurant entrance is on one side of the main dining room with a stage-like open kitchen. It shows everyone the wonderful cooking idea of traditional and innovative fusion cuisine from Piccolo Lago. Guests marvel at the forefront technology and a new generation of cookers, and can enjoy the chefs of "ancient" art.

The dining room is also decorated under the classical and modernist philosophy, embodying the fusion of two countries, from the elegant style of Louis XVI armchair, to the minimalist style of the European dreams of white and stone dining area and the mosaic walls and wooden panels. Private dining room is paved with red fir floor; an armchair with comfortable, quilted, oriental flavor is placed in the small alcove.

Meltino咖啡馆
Meltino Bar & Lounge

▶ 设计公司：克劳迪娅·科斯塔、洛夫工作室
▶ 面积：180m²
▶ Design Company: Cláudia Costa, Loff Atelier
▶ Area: 180m²

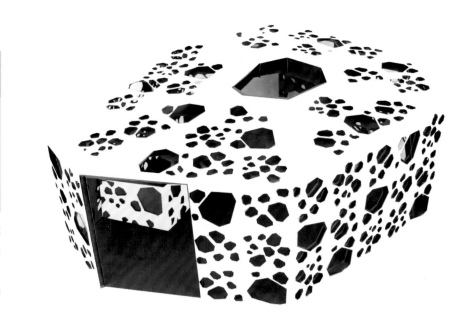

购物中心的地理位置对于咖啡馆而言或许并不是那么优越，但革新换代的咖啡理念，却洋溢在葡萄牙Meltino咖啡馆里。当所有的一切归功于本案设计，或许，优越性就此体现出来。

强烈的设计元素，刻意彰显着自己的存在。家具铺陈，似乎也成了咖啡新理念衍生的一部分。点滴浓香，点滴温馨，尽在空间内外，使休息大堂及

咖啡厅之间的阻隔悉数融化。

在不同的功能空间中,透明的气韵书写着空间的整齐划一。美食飘香,咖啡在手,一切令人心旷神怡。

孔状结构的渗透墙面映衬着咖啡豆的几何形状,光影交错,如九天飞舞。原本购物中心的布局已无影踪。这,就是本案设计的理念。当你举目四顾,墙体、屋顶、柜台到处是咖啡豆的化身。这,就是本案向世人的奉献。

早在本案还未实施之期,业主就已经寄予厚望,希望空间能以独特的气韵、十足的个性给客人留下经久不衰的印象。几经周折,一个不局限于原有商业布局、但量体之中另有量体的空间,以光感和透明的姿态跃然纸上。

咖啡的空间当然彰显咖啡的理念。几何化的咖啡豆如景幻化,或于墙上,或于天花。光影叠加,漫天飞舞。

空间两种划分,漫步区自然抬升,休息大堂和咖啡厅如月在卧。升降之间,彰显对比,半透明气质却缔结着连理。各种功能的家具自然铺陈。休息区咖啡美食尽在其中,咖啡区自然有极品咖啡相候。

几何形状的咖啡豆幻化着咖啡的浓香,飘逸在空间内外。行走在商场中,顺着咖啡的飘香,你会身不由己地来到咖啡空间。纤维板围合成一个个小小的空间,布置得十分自然。

双面墙

双面墙松木材料,中密度纤维板,体轻易建,比例适宜,强化着咖啡豆的概念。围合的空间天花自然延续主题,中央处微微隆起,倍添空间活力。包厢使用同样的材料,同样的洁白色泽。地板是油毯材质,棕色调。咖啡的纯洁和高贵品质尽在这棕、白二色的世界中深刻体现。

家具

陶质器皿和桌椅铺陈以细节营造美观、舒适的格调,引人注目。咖啡残渣制成的新材料的使用,颇令人思量,但却更容易引起共鸣。

照明设计

照明自然,或间照,或直照,平衡空间气氛。Logo标牌、橙色灯光、LED照明,彰显空间主人的身份和气质。

For a coffee shop, its location in the shopping center is not excellent. But the concept of innovation is filling the room. High praise should be there when refering to the design.

Having the strong presence of the design element, the furniture is the result of coffee derivatives. Aroma and warmth make it possibe for the connection between the resting hall and coffee shop.

Both are similar and transparent, although with different purposes. The lounge site has a relaxed atmosphere and offers gourmet products. The bar site is good for drinking coffee.

Permeable walls of cavernous structure set off the geometry of coffee beans, forming an image intertwining with light and shadow. The original layout of shopping center has disappeared. Walls, roof and counters and all that you can see have a shadow of coffee beans. That's the concept of this project.

High hopes were placed before the design, the owner hoping that it will have a unique atmosphere, and impressive personality. After several tries, the originally constricted layout becomes free with a clear feeling of brilliance and transparency.

The coffee shop of course shows the concept of coffee. Geometrical coffee beans is a changing shadow, a shadow that sometimes on the wall, sometimes at the ceiling. Light and shadows merge together.

There're two kinds of spaces. One is the raising waliking area. The other are the resting room and coffee shop sitting like the moon. They form contrast. Furniture of various functions are well arranged. For resting place, there's good food, while in the coffe shop there's the most excellent coffee.

Geometrical coffee beans give out an aroma of coffee in the air. Walking in this shopping center, guiding by the aroma, it's natural that you will come to this coffee shop that was divided into many small rooms enclosured by many fibreboards.

Double walls

The double walls are made up of pine wood and light fiberboards. It's not heavy and easy to build in an appropriate proportion, enhancing the concept of coffee beans. The enclosure space naturally continues the theme. The center slightly rise, providing more vitality. The floor covering is in linoleum while the box is in white. And the elegance manifests itself between white and brown color.

Furniture

The crockery and the tables were drawn in detail, aesthetics and comfort and attention. The furniture was covered with a unique and new material that is a derivative from the remnants of coffee. The furniture fulfills the space and appeals to sensorial feelings.

Lighting Design

The entire project was thought to create a balancing atmosphere through natural light or illumination. Logo brand, orange lamps and LED illumination show its status and personality.

天筑餐馆
El Japonez

- 设计公司：Cherem Serrano设计师事务所
- 摄影师：海梅·纳瓦罗、佩德罗
- 面积：290m²
- Design Company: Cherem Serrano Architects
- Photographer: Jaime Navarro, Pedro Hiriart
- Area: 290m²

本案设计独特，别具匠心。无立柱内里空间，气势横贯东西南北，光照充足。树木掩映，绿植如若天赐。空间绿意盎然，而无盆栽之娇柔、小气。入内就餐，翡翠入眼，如在湖边、山涧、溪边小炊。

塑料地板，任其挥发着传统日本和室的榻榻米质感。

空间中独立支柱，给人以单独建筑之感。7个柱体装饰从天花处直下，有立柱之气度，却无其之实体。无法触地的诙谐形态，让人忍俊不禁。两个不同的环境在同一情境下营造。嵌入式的立柱墙设计，当然铸就了空间的贯通，光影的自由流动。

立柱墙体、天花饰以10cm×10cm壁面，延续着空间的木香质感，彰显着设计的"自然"主题。

酒吧区的上层空间，立柱墙开设孔洞，嵌以照明系统，弱化了墙体产生

的厚重感,产生了一种神秘感,使人不禁想一探究竟。 酒吧区的后方隐藏着楼梯,可通往休息室。远远望去,恰如不透明的玻璃盒体隐藏在另一个木盒中。那里的环境、气氛同样诙谐,原本分离的男、女卫生间却无意中弱化着彼此之间的距离。

The project is a unique design. A large open space, full of light, with no virtually columns, is planted with green trees in four angles, with adequate daylight. There's no delicacy like the potted plants. When having dinner here, you will have a feeling of dining at the side of lake and stream, or among the hills.

The floor of the restaurant is covered by a plastic carpet that evokes the texture of the tatami of Japanese Architecture.

The scarcity of columns is evident: only one column is clearly present in the space, which creates the impression that this long stretch is supported by only one structural element. There are other seven elements that playfully pretend to be columns but never touch the ground: they emerge from soffit and have a specific role: the creation of two different environments within the same atmosphere. The rest of the columns have been hidden so as to avoid interrupting the flow of light and space.

It continues the wooden texture and shows the theme of nature through the objects of 10cm×10cm decorated on the wall and ceiling.

Over the bar, the stud wall shows some cavities, which are illuminated, making it appear less heavy and revealing that something happens behind them. A staircase hidden behind the bar leads to the rest rooms, which feature an opaque glass box contained in another wooden box. There, the environment is milder, and it playfully pretends to minimize the sense of separation between the men's and the women's rest rooms.

Salon Des Saluts酒吧
Salon Des Saluts

- 设计公司：缘之土设计
- 设计师：其卡拉琴
- 摄影师：矢野敏行
- 面积：53.51m²
- Design Company: Natural Earth Inc.
- Designer: Chikara Ohno
- Photographer: Toshiyuki Yano
- Area: 53.51m²

本案位于东京，直面Seijouki街，定位为酒吧。

除考量空间内外，功能界限、空间透明两不耽误。基地虽小，但视线通畅。四个玻璃小方体依次排列，整齐划一。其中一个用作包房，另外三个或植以单株橄榄，或用作路径，或连接阳台内外。动线行进曲折复杂，内外界限因此合二为一。各方体空间角落皆以曲线处理，视觉亦真亦假；内外划分时而模糊，时而透明，令其空间层次因而丰富、变幻。

而玻璃墙、地砖两相映照，曲线如波涛般前行。芦笋、迷迭香、常春藤等绿色植物依势而行，倏忽不定的内外界限进而得以强化。

置身于酒吧之内，却如游离于东京城市风情；行走于人群之中，却如小憩于酒吧之内。这就是设计精华所在。

We designed a small wine bistro fronting straight onto the Seijouki Street, Nishiazabu, Tokyo.

Except the regard between the outside and the inside space, we are considering the value of ambiguous boundary and deep transparency. Although the site is very small, it gets a good vision. We decided to put 4 small glass boxes by order into an opening of the building that crosses over the outline. One box is used for private room inside, while the other three boxes are used for a court planted with single olive tree, a path and a terrace. It's difficult to know where a real boundary between the inside and the outside is, because this line runs very complicatedly. Besides, every corner of boxes is processed in curves. For the division of the inside and the outside, sometimes it blurs, and sometimes it gets good transparency. A sense of levels fills the space.

The glass walls echoes with the ground tiles. Its curves move like the wave. Asparagus, rosemary and ivy and so on are planted along the curve. There's no clear boundary due to its frequent change.

We hope the shop we designed becomes a kind of public amenity and makes a good experience not only for customers, but for people walking along the street.

Shook!餐厅
Shook!

- 设计公司：Orbit Design
- 业主：杨忠礼酒店集团
- Design Company: Orbit Design
- Owner: YTL Hotels

　　YTL Hotels（杨忠礼酒店集团）旗下的上海Shook!餐厅位于上海斯沃琪和平饭店艺术中心，这里曾经是汇中饭店所在地。餐厅所在地的原汇中饭店大楼是一幢兴建于1906年的文艺复兴风格的大楼，历经一个多世纪，依然风姿卓绝。大楼位于南京东路与外滩的交界处，从这里可以俯瞰黄浦江和浦东的天际线。

　　Shook!享誉世界，其成功始于马来西亚吉隆坡杨忠礼旗下Starhill Gallery里的Shook!餐厅。上海Shook!是马来西亚首都吉隆坡之外Shook!的首家分店。餐厅拥有一个昵称为"法拉利"的梦幻厨房，在那里由顶级名厨掌勺，烹饪出一道道让人垂涎欲滴的美味佳肴。

　　通往餐厅的走道两侧是布置精美的酒窖，展示着世界闻名的葡萄酒和香槟。酒窖的压窖之宝乃是两款波尔多精品系列酒，即白马庄酒（Chateau Cheval Blanc）和狄康堡酒（Chateau d'Yquem）。每个系列均涵盖了从1945到2010年各个年份的产品。

　　客人可以先去Time Bar小酌，开始一段美妙的旅途。酒吧用深色木材和黑石装饰，点缀着奶黄色的皮革。这里的陈设与氛围相契无间，尽显典雅。长长的玻璃窗把壮丽的黄浦江景呈现在客人面前。用餐区精美的纹理

与华贵的织物形成古今碰撞的鲜明对比,流露出旧日欧洲流行的中国风(Chinoiserie)印记。

餐厅的露台于2011年春季开放,这里的设计简单至极。白色的长沙发,精心点缀着黑红两色花纹,营造出绝佳的酒吧氛围,它将成为上海这个国际大都市精英人士难忘的欢聚之地。

五楼邻近上海Shook!餐厅的是高雅的多媒体宴会厅"艺廊(The Gallery)"。这是一个举办音乐会、放映电影、举办新闻发布会和私人宴请的绝佳地点。大楼屋顶露台上两个标志性的带穹顶的塔"汇中阁(Corner Tower)"和"宝珀阁(Blancpain Room)"均有高高的天花板,其典雅的装饰风格非常适合举办私密的烛光晚宴。位于六楼的"美庐厅(Chiang-Soong Room)"见证了上海的传奇历史,堪称沪上一景。美庐厅现已修葺一新,恢复了旧时的典雅。这里曾是1909年万国禁烟会的举办地,蒋介石和宋美龄也在这里举行了订婚仪式。

杨忠礼酒店集团选择在原汇中饭店开设上海Shook!餐厅,可谓眼光独到。外滩东西碰撞,新旧交融,客人在此,不禁产生时空交错之感。

Located in the art center of Shanghai Swatch Peace Hotel, formerly Huizhong Hotel that is built in 1906 with Renaissance style, Shook! Shanghai belongs to YTL Hotels. Over a century later, its elegance still appeals to us. It overlooks the horizon between Huangpu River and Pudong District, situating on the juncture of the eastern Nanjing Road and Waitan in Shanghai.

Shook! Shanghai—the first outpost out of the Malaysian capital, Kuala Lumpur, has won its worldwide reputation and its secret of success comes from the Shook! of Starhill Gallery belonging to YTL Hotels. A show kitchen—dubbed 'Ferrari' which means a dreamy kitchen, is an excellent kitchen with many excellent chefs, which of course leads to many tasteful dishes.

Lining walls on both sides of the hallway leading to the restaurant are the beautifully designed wine cellars, showcasing some of the world's finest wines and champagnes. The centerpiece of the wine cellar is made up of two rare vertical collections of Bordeaux—Chateau Cheval Blanc and Chateau d'Yquem. Each collection offers vintages from 1945 to 2010.

You can begin at Time Bar and enjoy yourself by drinking something. It's a bar accented with dark wood and black stone, decorated with cream-colored leather. It is an excellent balance between furnishings and the setting. The view of spectacular Huangpu River meets your eyes thanks to the long glass windows. In the dining area, wrought textures and sumptuous fabrics provide a contrast between the past and present, showing the once popular Chinoiserie in ancient Europe.

The terrace of the restaurant will be open in the spring of 2011. Its design is very simple with white couches dotted with black and red patterns, making an excellent atmosphere. It's true that this restaurant will be one of the most attractive places in Shanghai.

Adjacent to Shook! Shanghai on the fifth floor, the stylish multi-media banqueting hall "The Gallery" is the perfect venue for concerts, private film, press conferences and private receptions. The hotel's two defining domed towers on The Swatch Art Peace Hotel Terrace, Corner Tower and Blancpain Room, both with high-ceilinged elegant décor are ideal for intimate candlelight dinners. On the sixth floor is the Chiang-Soong Room, witnessing the legend history of Shanghai, as one of the attractions in Shanghai. It has been restored to its original elegance. The Chiang-Soong Room is the very same venue where the First International Opium Conference was held in 1909. It is also the place where Chiang Kai-Shek and Soong Mei Ling of the storied Soong Sisters celebrated their engagement.

It's a wise and unusual decision for YTL Hotels to open its shop, Shook! Shanghai. It meets the point of combination of the oriental and occidental culture, and of the old and the new, giving rise to a sense of time and space meeting together.

Affranchir餐厅
Affranchir

- 设计公司：Cherem Serrano设计师事务所
- 设计师：衫山敦彦
- Design Company: Cherem Serrano Architects
- Designer: Atsuhiko Sugiyama

外表简洁、装饰少、形体模块简单，是日本长期以来的设计风格。本案外观宁静、简朴，给人一种优雅的感觉。

走近Affranchir，首先邂逅的是入口处两排错落有致的棕榈树，热带风情悠然呈现。从树间透露出来的两边褚红的砖墙和墙体交接处的一排圆灯，巧妙地诠释着建筑的日式风格。

Affranchir为来访者带来悠闲舒适的度假体验。接待厅设计别出心裁，沿墙一排矮书柜整齐地摆放着各类书籍，令你好像走进自家书房一样轻松。进入就餐大厅，你会同样感觉轻松自在。空间虽然有些狭窄，但是主厅双层挑高的设计，让你在空间中顿生舒展开阔之感。沿楼梯信步而下，曲线曼妙的铁艺扶手，犹如波浪一般清新可感。行至底层，抬头回望，只见二楼一角，由铁丝围成的一角书架，错落地放着一些图书做装点，仿佛提醒你要以阅读的心态看待人生的波浪起伏。

酒吧台处有各式生活小物件轻松混搭。马赛克地毯、小块红砖地板、线条简约的咖啡色材质墙面、墙上的几本旅游相册，在温暖的墙面交接处摆放一个长方形书架和一张书桌，把精神层面的文化引入到就餐环境中，让你在享受美食之前先来一份精神粮食作为点心。

Affranchir的水边吧台也极具吸引力。在长方形的水池内，镶嵌着一方小

竹林，郁郁葱葱的修竹向天空自由伸展着，带来大自然的勃勃生气。水池旁还有几堵矮墙，特别设计的流水从墙中跌落，犹如人造小瀑布。坐在水池边的吧台前，喝着清凉的饮品，赏着竹林青翠，听着流水哗哗，尘世的喧闹仿佛已丢在一边，让人尽享不受干扰的宁静和悠然。

Concise appearance with few decorations is the traditional Japanese style of design and most of it shows an air of elegance.

Getting close to Affranchir, firstly you'll encounter palm trees properly distributed on both sides of the entrance, showing the tropical feeling from the red brick wall. There is a row of round lamps at the combining points of the wall, expressing the Japanese style.

What Affranchir brings to customers are leisure and coziness. The reception hall is designed uniquely. There is a row of short bookshelves along the wall with kinds of books placed orderly on them, which make you feel as casual as in your own study. The dining hall coming after is as leisure and cozy as the reception hall. Although it is kind of small, the two-story high ceiling eases the hearts of people here. When coming down the stairs leisurely, the winding iron handrail is so lovely. Looking up the stairs you can see at the corner of the second floor a bookshelf made of iron wires decorated with some books, which seems to remind you of state of the ups and downs of life that you'd better live with the attitude of reading.

In the bar area, there are many little objects that are casually mixing and matching together. There is mosaic carpet, red brick floor and streamline brown wall with some books about the guide for traveling, and at the intersection of the wall there is a rectangle book shelve and a desk, which take the spirit into the environment, as if they were the dim sum in spirit.

The bar beside the pond is also very attractive. The green bamboos in the middle of the rectangle pond are growing freely into the sky, bringing in the vitality of nature. Some special designed short walls around the pond letting the water flow down like a small waterfall. Sitting by the side of the pond, drinking the cool beverage, enjoying the bamboos and listening to the music, all make you appreciate the quietness and leisure of life without the trouble and noises of the outside world.

The Japonais餐厅
The Japonais

设计公司：Cherem Serrano设计师事务所
设计师：衫山敦彦
Design Company: Cherem Serrano Architects
Designer: Atsuhiko Sugiyama

棕榈树、马赛克拼花、红砖模块墙、烧砖圆柱和大地色系在空间里到处蔓延。东西方交融的文化和美感在Japonais的空间里一一得以呈现。

入口处简约的设计带给来访者一种休闲之感。高低错落的几棵棕榈树装点着红色砖墙，靠近门处，红砖横竖拼贴出编织纹样，内嵌黑白马赛克的门框，东西文化的交融在入门处可见一斑。

踏着门前红绿相间的地毯，这里没有繁复的装饰，没有琳琅满目的点缀。廊道两旁的墙壁尽显日式风格的简朴线条，两旁对称悬挂的灯笼折射出的火红给砖块堆砌的墙面增添了几许热情。玻璃门扇上的黄铜拉手流露出历史的沧桑感。

红砖拼贴的墙身图案，延续到空间的每个角落，无论是餐厅正厅还是小包厢抑或吧台，都使内外空间保持着整体的一贯性。弧形吧台旁，红砖拼贴的折扇式墙身，与前面的木栅门材料对比，丰富了空间的层次感；就餐区，红砖拼贴出的壁炉图形，充满了田园的休闲气氛；还有礼堂里，红色砖墙与红灰火烧砖拼贴的柱子共同构筑出一个温馨暖和的空间，为婚礼的举行增加浪漫之感。

怎样才能将生活的娱乐感与休闲感带入空间里？Japonais选择将"赌城"的一些趣味元素融入空间设计之中。弧形吧台的桌面被设计成数"赌桌"样式，颜色深浅相间的小格里标注着各式下注的数字，营造出一种休闲娱乐的氛围。连休息的包厢也借用了赌城的元素进行装点，宽厚的大沙发，松软得让人陷在里面不想出来。沙发侧面抽象的图案透露出"赌城"的诱惑，这或许是设计师给大家开的一个小小玩笑。

轻松、愉悦的氛围充溢在空间的各个角落。不同组合式的坐席，如团座式、卡座式、包厢式，以及长条沙发的大聚合，让来宾可以各得其乐，享受自由交谈的乐趣。

Palm trees, mosaic flowers, red brick wall, bake-brick column and earth hues can be seen in the whole space. The beauty out of eastern and western culture appears in The Japonais.

The simple design for the entrance brings visitors a feeling of leisure. Some palm trees decorate the red brick wall. Close to the door, there are patterns weaved by red bricks while the doorframe is imbedded with black and white mosaic. The blend of eastern and western culture shows here at the entrance.

The green and red carpet at the entrance is clear without complicated and colorful decorations, while the walls on both sides of the corridor are built in the Japanese style with simple lines and the red shadows refracted by the lanterns hung symmetrically make the brick wall full of some kind of passion. The brass door knocker on the glass door expresses a sense of old history.

The patterns in the collage of red bricks fill every corner in the dining room and small boxes or bars, keeping the unity of the whole space. Beside the arch bar desk, the arc wall of red bricks contrasted with the wood fence in front of it, which enrich the sense of space layers. In the dining area, the pattern of fireplace made of red bricks possesses the quality of leisure. There is a small auditorium of warmth and romance in which you can hold a small romantic wedding.

How to bring the recreation and leisure of life into the space? Here, Japonais has tried some interesting elements of casino. The arc desks in the bar area are designed into the type of casinos with small cells of deep and shallow colors. Each small cell has a number on it, which creates a sense of leisure and recreation. The huge sofas in the rest room make one who sets in it not want to get out of it. Even the abstract patterns on the sides of the sofa show the lure of casinos, which may be a small joke the designer makes for us.

The air of leisure and happiness is everywhere in this space. The different combinations of seats allow customers to enjoy themselves through free chatting.

大卫迈尔斯咖啡厅
Davis Coffee

- 设计公司：野村有限公司
- 设计师：柳小坂、中村
- 摄影：Nacása & Partners公司
- 面积：115m²
- Design Company: Nomura Ltd.
- Designer: Yakosaka, Nakamura
- Photograph: Nacása & Partners Company
- Area: 115m²

本案地理位置优越，位于银座中心三越百货公司六楼。邻近专柜销售男装、雪茄、手表、眼镜等高档商品。业主大卫来自阳光海岸美国加州，是专业厨师，向往简约、豪华。本案作为投石问路之作，以其考究的设计，大方的室内铺陈引起了世人的广泛关注。中心区预制木制材料、悬挂饰品引人注目。石膏墙简约、大方、自然，使人如浴春风。灯光漫射，加上正前方格栅的作用，使墙体斑驳，光影闪动。本案设计跳脱出日式空间设计的传统，令人耳目一新，彰显着空间的深度与优雅。

Situating on the sixth floor of Sanyue Shop in Ginza Center, this project has a good place. Neighboring shops are high-end stores selling men's clothes, cigar, watches and eyeglasses. The owner of this project, a professional chef comes from California, the USA, He likes simplicity and luxury. This project, a trial project, attracts people's attention because of its excellent design and good arrangement for its interior. The custom-made wooden materials and suspended decorations have won many people's admiration. The simple, large and natural plaster walls giving out a sense of nature, like a spring breeze. The spreading light and grids in the front and the mottled walls make the shade dance. In a word, this design is very new, absolutely not a slave to the traditional Japanese design, showing its depth and elegance.

图书在版编目（CIP）数据

宴遇 / 黄滢主编. -- 南京：江苏科学技术出版社，2014.4
 ISBN 978-7-5537-2960-2

Ⅰ．①宴… Ⅱ．①黄… Ⅲ．①餐馆－室内装饰设计－图集 Ⅳ．①TU247.3-64

中国版本图书馆CIP数据核字(2014)第045587号

宴遇

主　　　编	黄　滢
项 目 策 划	凤凰空间
责 任 编 辑	刘屹立
出 版 发 行	凤凰出版传媒股份有限公司
	江苏科学技术出版社
出版社地址	南京市湖南路1号A楼，邮编：210009
出版社网址	http://www.pspress.cn
总 　经　 销	天津凤凰空间文化传媒有限公司
总经销网址	http://www.ifengspace.cn
经 　　　销	全国新华书店
印 　　　刷	北京博海升彩色印刷有限公司
开　　　本	1020毫米×1440毫米　1/16
印　　　张	20
字　　　数	160 000
版　　　次	2014年4月第1版
印　　　次	2014年4月第1次印刷
标 准 书 号	ISBN 978-7-5537-2960-2
定　　　价	428.00元

图书如有印装质量问题，可随时向销售部调换（电话：022-87893668）。